CRC SERIES IN CHROMATOGRAPHY

Gunter Zweig and Joseph Sherma
Editors-in-Chief

CRC Handbook of Chromatography: Polymers

Volume I

Editors

Charles G. Smith
Analytical Laboratories
Dow Chemical USA
Midland, Michigan

Norman E. Skelly, Ph.D.
Analytical Laboratories
Dow Chemical USA
Midland, Michigan

Richard A. Solomon
Analytical Laboratories
Dow Chemical USA
Midland, Michigan

Carl D. Chow
Analytical Laboratories
Dow Chemical USA
Midland, Michigan

CRC Series in Chromatography

Editors-in-Chief

Gunter Zweig, Ph.D.
School of Public Health
University of California
Berkeley, California

Joseph Sherma, Ph.D.
Professor of Chemistry
Lafayette College
Easton, Pennsylvania

CRC Press, Inc.
Boca Raton, Florida

Library of Congress Cataloging in Publication Data

Main entry under title:

Polymers.
(CRC handbook of chromatography)
Bibliography: p.
Includes index.
1.Polymers and polymerization —Analysis.
2. Chromatographic analysis. I. Smith,
Charles G., 1940- II.Series.
QD381.P6124 547.7'046 81-10097
ISBN 0-8493-3073-4 AACR2

Direct all inquiries to CRC Press, Inc., 2000 Corporate Blvd., N.W., Boca Raton, Florida, 33431.

Second Printing, 1986

International Standard Book Number 0-8493-3073-4

Library of Congress Card Number 81-10097
Printed in the United States

CRC HANDBOOK OF CHROMATOGRAPHY

SERIES PREFACE

The present volume on the chromatography of synthetic polymers, edited by Charles G. Smith, Richard A. Solomon, Norman E. Skelly, and Carl D. Chow, of the Dow Chemical Company, is an important addition to the series of *Handbooks of Chromatography* begun in 1972 under the aegis of CRC Press. Recently published volumes of this series have covered the subjects of Drugs and Carbohydrates. The analysis and characterization of polymers is greatly aided by various chromatographic methods, and the authors have very ably collected, organized, and edited the most important chromatographic information for polymers, monomers, and oligomers in this volume. This book, like others in the series, consists in the first part of detailed tables of chromatographic data for the important techniques such as size exlcusion (gel permeation) chromatography, gas chromatography, thin layer chromatography, and liquid chromatography. The second part of the book is descriptive, giving readers fundamental and practical information on the various techniques especially applicable to polymers and related compounds. This volume is limited to the chromatography of synthetic polymers; natural polymers such as polysaccharides and polynucleotides and nucleic acids are covered elsewhere in the series. As in the past, we ask our faithful readers to inform us of any shortcomings in the book and to call our attention to inadvertent mistakes that invariably go undetected even in the most carefully proofread manuscript.

Gunter Zweig
Joseph Sherma
May, 1982

PREFACE

Volume I of the *Handbook of Chromatography* contains tables of chromatographic data for over 12,000 compounds. The broad scope of these tables includes steroids, drugs, and metal ions in addition to a large number of organic compounds. Volume II presents the theory and practice of gas, liquid-column, thin-layer and paper chromatography. These volumes provide the background for subsequent sections on specific chromatographic applications.

This Volume presents data on the chromatographic analysis of polymers and polymer related compounds. Supplemental theory subsections for gas, liquid-column and thin-layer chromatography describe column and detection systems new to their respective fields since publication of Volume II. This Volume also describes the theory and practice of two specialized techniques for polymer analysis — pyrolysis-gas chromatography and size exclusion chromatography.

This Volume includes tables of chromatographic data abstracted from 1970 to 1980 literature references. Compounds tabulated include residual monomers, plasticizers, additives, antioxidants, and products from pyrolysis of a broad range of polymers. Where applicable, tables include specialized techniques for extraction of additives from the polymer matrix prior to chromatographic analysis.

The Practical Applications subsection provides the reader with a list of commercial suppliers of column packings and packed columns or thin-layer plates.

Finally, the editors wish to acknowledge the contribution of John Cobler who initiated this compilation of polymer chromatographic data 2 years ago but was unable to participate in completion of the manuscript.

Charles G. Smith

THE EDITORS-IN-CHIEF

Gunter Zweig, Ph.D., received his undergraduate and graduate training at the University of Maryland, College Park, where he was awarded the Ph.D. in biochemistry in 1952. Two years following his graduation, Dr. Zweig was affiliated with the late R. J. Block, pioneer in paper chromatography of amino acids. Zweig, Block, and Le Strange wrote one of the first books on paper chromatography which was published in 1952 by Academic Press and went into three editions, the last one authored by Gunter Zweig and Dr. Joe Sherma, the co-Editor-in-Chief of this series. *Paper Chromatography* (1952) was also translated into Russian.

From 1953 till 1957, Dr. Zweig was research biochemist at the C. F. Kettering Foundation, Antioch College, Yellow Springs, Ohio, where he pursued research on the path of carbon and sulfur in plants, using the then newly developed techniques of autoradiography and paper chromatography. From 1957 till 1965, Dr. Zweig served as lecturer and chemist, University of California, Davis and worked on analytical methods for pesticide residues, mainly by chromatographic techniques. In 1965, Dr. Zweig became Director of Life Sciences, Syracuse University Research Corporation, New York (research on environmental pollution), and in 1973 he became Chief, Environmental Fate Branch, Environmental Protection Agency (EPA) in Washington, D.C.

During his government career, Dr. Zweig continued his scientific writing and editing. Among his works are (many in collaboration with Dr. Sherma) the now 11-volume series on *Analytical Methods for Pesticides and Plant Growth Regulators* (published by Academic Press); the pesticide book series for CRC Press; co-editor of *Journal of Toxicology and Environmental Health;* co-author of basic review on paper and thin-layer chromatography for *Analytical Chemistry* from 1968 to 1980; co-author of applied chromatography review on pesticide analysis for *Analytical Chemistry,* beginning in 1981.

Among the scientific honors awarded to Dr. Zweig during his distinguished career are the Wiley Award in 1977, Rothschild Fellowship to the Weizmann Institute in 1963/64; the Bronze Medal by the EPA in 1980.

Dr. Zweig has authored or co-authored over 75 scientific papers on diverse subjects in chromatography and biochemistry, besides being the holder of three U.S. patents.

At the present time (1980/81), Dr. Zweig is Visiting Scholar in the School of Public Health, University of California, Berkeley, where he is doing research on farmworker safety as related to pesticide exposure.

Joseph Sherma, Ph.D., received a B.S. in chemistry from Upsala College, East Orange, N.J. in 1955 and a Ph.D. in analytical chemistry from Rutgers University in 1958. His thesis research in ion exchange chromatography was under the direction of the late William Rieman III. Dr. Sherma joined the faculty of Lafayette College in September 1958, and is presently full professor there in charge of two courses in analytical chemistry. At Lafayette he has continued research in chromatography and has additionally worked a total of 12 summers in the field with Harold Strain at the Argonne National Laboratory, Illinois, James Fritz at Iowa State University, Ames, Gunter Zweig at Syracuse University Research Corporation, New York, Joseph Touchstone at the Hospital of the University of Pennsylvania, Philadelphia, Brian Bidlingmeyer at Waters Associates, Framingham, Mass., and Thomas Beesley at Whatman Inc., Clifton, N.J.

Dr. Sherma and Dr. Zweig [who is now with U.S. Environmental Protection Agency (EPA)] co-authored Volumes I and II of the *CRC Handbook of Chromatography,* a book on paper chromatography, and 6 volumes of the series *Analytical Methods for Pesticides and Plant Growth Regulators.* Other books in the pesticide series and further volumes of the *CRC Handbook of Chromatography* are being edited with Dr. Zweig, and Dr. Sherma will co-author the handbook on pesticide chromatography. A book on quantitative TLC (published by Wiley-Interscience, New York) was edited jointly with Dr. Touchstone. Dr. Sherma has been co-author of 7 biennial reviews of liquid chromatography (1968 to 1980) and the 1981 review of pesticide analysis for the journal *Analytical Chemistry.* Dr. Sherma has authored major invited chapters and review papers on chromatography and pesticides in *Chromatographic Reviews* (analysis of fungicides), *Advances in Chromatography* (analysis of fungicides), *Advances in Chromatography* (analysis of nonpesticide pollutants), Heftmann's *Chromatography* (chromatography of pesticides), Race's *Laboratory Medicine* (chromatography in clinical analysis), *Food Analysis: Principles and Techniques* (TLC for food analysis), *Treatise on Analytical Chemistry* (paper and thin layer chromatography), and *CRC Critical Reviews in Analytical Chemistry* (pesticide residue analysis). A general book on thin layer chromatography co-authored by Dr. Sherma is now in press at Marcel Dekker.

Dr. Sherma spent 6 months in 1972 on sabbatical leave at the EPA Perrine Primate Laboratory, Perrine, Fla., with Dr. T. M. Shafik, and two additional summers (1975, 1976) at the U.S. Department of Agriculture (USDA) in Beltsville, Md., with Melvin Getz doing research on pesticide residue analysis methods development. He spent 3 months in 1979 on sabbatical leave with Dr. Touchstone developing clinical analytical methods. A total of more than 200 papers, books, book chapters, and oral presentations concerned with column, paper, and thin layer chromatography of metal ions, plant pigments, and other organic and biological compounds; the chromatographic analysis of pesticides; and the history of chromatography have been authored by Dr. Sherma, many in collaboration with various co-workers and students. His major research area at Lafayette is currently quantitative TLC (densitometry), applied mainly to clinical analysis and pesticide residue determinations.

Dr. Sherma has written an analytical quality control manual for pesticide analysis under contract with the U.S. EPA and has revised this and the EPA Pesticide Analytical Methods Manual under a 4-year contract (EPA) jointly with Dr. M. Beroza of the Association of Official Analytical Chemists (AOAC). Dr. Sherma has also written an instrumental analysis quality assurance manual and other analytical reports for the U.S. Consumer Product Safety Commission, and is currently preparing a manual on the analysis of food additives for the U.S. Food and Drug Administration, both of these projects also in collaboration with Dr. Beroza of the AOAC.

Dr. Sherma taught the first prototype short course on pesticide analysis with Henry Enos of the EPA for the Center for Professional Advancement. He is editor of the Kontes TLC quarterly newsletter and also teaches short courses on TLC for Kontes and the Center for Professional Advancement. He is a consultant for several industrial companies and federal agencies on chemical analysis and chromatography and regularly referees papers for analytical journals and research proposals for government agencies.

Dr. Sherma has received two awards for superior teaching at Lafayette College and the 1979 Distinguished Alumnus Award from Upsala College for outstanding achievements as an educator, researcher, author, and editor. He is a member of the American Chemical Society, Sigma Xi, Phi Lambda Upsilon, Society for Applied Spectroscopy, and the American Institute of Chemists.

THE EDITORS

Charles G. Smith works with the Polymer Analysis Group of the Analytical Laboratories, Michigan Division of the Dow Chemical Company®. He received his B.S. degree from Alleghany College and his M.S. in chemistry from the University of Michigan. Following graduation, he worked three years with the Chemical Division of PPG Industries before joining Dow in 1967.

After 4 years of methods development in the Organic and Agricultural Products Groups within the Analyical Laboratories, Mr. Smith transferred to the Polymer Analysis Group. His expertise over the years has been concentrated on chromatographic separations (gas and liquid) and applications to polymer systems. The last 4 years were spent developing pyrolysis-gas chromatographic capabilities for polymer characterization.

He is chairman of two sections of the American Society for Testing and Materials Committee on analytical method for plastics. Mr. Smith is author or co-author of three publications and numerous Dow® analytical methods.

Dr. Norman E. Skelly has worked on a wide variety of problems in his career at the Dow Chemical Company®. He entered the University of Iowa in 1951 and received his M.S. and Ph.D. degrees in analytical chemistry in 1953 and 1955, respectively. Upon graduation, he joined the Dow Chemical Company®, Michigan Division, where he has the present title of Associate Scientist. His areas of interest include liquid and thin-layer chromatography, electrophoresis and ultraviolet spectrophotometry as they pertain to the solution of organic analytical problems. He is the author or co-author of 25 publications and over 150 Dow® analytical methods.

Carl D. Chow is with the Polymer Analysis Group of the Analytical Laboratories, Michigan Division of the Dow Chemical Company®. He received his B.S. in organic chemistry from the National Taiwan University and his M.S. in polymer chemistry from North Dakota State University. He joined the Physical Research Laboratory of Dow® in 1965 and worked on polymerization kinetics, foam technology, and reinforced plastics. He transferred to the Polymer Analysis Group in 1967. His specialties are molecular weight measurements, gel permeation chromatography (GPC), liquid chromatography, and thermal analysis. He has published in these areas and has lectured on GPC at local ACS sections and universities. He has been one of two review authors for the Application Reviews issue of Analytical Chemistry on Analysis of High Polymers since 1975.

Richard A. Solomon has specialized in the field of gas chromatography during his career at the Dow Chemical Company®. He received his B.S. from Westminster College, New Wilmington, Pa. and his M.S. from Pennsylvania State University in 1955.

He is the author or co-author of five publications and over 200 Dow® analytical methods.

ADVISORY BOARD

TABLE OF CONTENTS

Section I
Principles and Techniques

I.1 GAS CHROMATOGRAPHY

The basic apparatus and theory of gas chromatography are discussed in *Section A* of the *CRC Handbook Series in Chromatography.*[1] The reader is referred to this same section for details regarding column selection, qualitative and quantitative analysis by conventional gas chromatography. Specific requirements to prepare polymer samples for gas chromatographic analysis (i.e., extraction, derivatization, etc.) are summarized in the tables of Section IV.I in this volume.

Headspace Gas Chromatography

Gas chromatographic headspace analysis is an indirect method for the determination of volatile constituents in liquids or solids by gas chromatographic analysis of the vapor phase which is in thermodynamic equilibrium with the sample in a closed system. Theory and practice of this technique are detailed in books.[2-3] Kolb[4] reviews the applications of automated headspace analysis.

In practice, the sample to be analyzed is placed into a glass vial and sealed with a rubber septum. The vial is heated in a constant temperature bath until equilibrium between sample and vapor phase is established for each volatile component. Each component has a unique equilibrium constant or partition coefficient which must be determined experimentally to obtain quantitative analysis. Matrix effects play an important role in determination of these constants. Once equilibrated, a known aliquot of the vapor phase is transferred to the chromatographic column for separation. This transfer is achieved using a gas-tight syringe or, in the case of automatic instruments, an electropneumatic dosing system. With careful calibration, the composition of the vapor phase obtained from the gas chromatogram is related to the volatile content of the original solid or liquid sample. Backflush systems are often employed to eliminate solvent peaks and interferences.

Headspace analysis has achieved considerable importance in the polymer field where the efficient resolution available and the high sensitivity and specificity of gas chromatographic detectors can be exploited. The applications of headspace analysis include the determination of residual solvents in solid chemicals and packaging materials, residual monomers in polymers and olfactory substances in plastics and polymer dispersions.[5,6] The advantages of this method compared with conventional manipulation of samples are simple sample preparation, minimal analysis time, high sensitivity, and minimal contamination of the column with high boiling or nonvolatile materials.

Detectors

Details of thermal conductivity and flame ionization detector theory are reviewed in an earlier section of this work.[1] The principles of separation for thermionic nitrogen-phosphorus detectors applicable in polymer analysis and other areas are summarized in the following paragraphs.

The thermionic phosphorus-nitrogen detector (PND) is specific for phosphorus and nitrogen containing compounds. An alkali source, connected to negative potential, is situated in the flame of a flame ionization detector. In the presence of phosphorus or nitrogen compounds, a thermionic reaction takes place with the release of negative ions. These are captured by the positive collector electrode giving rise to a detector signal which is supplied to the FID amplifier.

The alkali source consists of a glass bead made of a special rubidium glass fused onto a platinum wire. As the alkali (Rb) is initially present as a nonvolatile rubidium silicate and as conversion to atomic alkali takes place only during the thermionic process, the loss of alkali ions is extremely small, with the result that the detector performance remains constant over a long period of time.

REFERENCES

1. **Sherma, J.,** Gas chromatography principles and techniques, *in Handbook Series in Chromatography,* Vol. II, Zweig, G. and Sherma, J., Eds., CRC Press, Boca Raton, Fla., 1972, 7.
2. **Kolb, B.,** *Applied Headspace Gas Chromatography,* Heyden and Sons, Inc., Philadelphia, 1980.
3. **Hachenberg, H. and Schmidt, A. P.,** *Gas Chromatographic Headspace Analysis,* Heyden and Sons, Inc., Philadelphia, 1977.
4. **Kolb, B.,** *J. Chromatogr.,* 122, 533, 1976.
5. **Steichen, R.,** *Anal. Chem.,* 48, 1398, 1976.
6. **Lattimer, R. P. and Pausch, J. B.,** *Am. Lab.,* 12, 80, 1980.

I.2 PYROLYSIS-GAS CHROMATOGRAPHY

Pyrolysis normally involves the thermal decomposition of organic molecules in the absence of oxygen. Pyrolysis products or fragments generally relate to the structure of the parent compound. Although these decomposition fragments are usually smaller molecules than the parent compound, pyrolysis can result in isomerization or formation of higher molecular weight fragments.

Pyrolysis-gas chromatography (P-GC) combines this thermal decomposition step with gas chromatographic (GC) separation of the pyrolysis fragments. Pyrolysis-gas chromatography-mass spectrometry (P-GC-MS) or direct pyrolysis-mass spectrometry (P-MS) are used to identify pyrolysis fragments. Although theoretically applicable to any organic compound with insufficient volatility for GC separation alone, this text is devoted mostly to P-GC as it applies to polymer analysis. Procedures utilizing mass spectrometric detection are referenced, but details of these detector systems are beyond the scope of this work. Details on the theory and practice of gas chromatography and detectors are described in earlier volumes[1] and Section I.1 of this volume. This section is devoted solely to pyrolysis theory, instrumentation, and applications. Irwin[2] recently reviewed the field of analytical pyrolysis including a bibliography of review articles and an earlier comprehensive review of the technique was presented by Levy[3]. Books describing theory and applications of analytical pyrolysis are also available.[4,5]

Theory

Thermal energy is applied to the polymer sample in the absence of oxygen. For convenience, pyrolysis is frequently performed in the gas chromatographic carrier gas, i.e., helium or nitrogen. The thermal energy applied to the polymeric sample exceeds the molecular bond energies resulting in fragmentation. Mechanisms of fragmentation processes for polymers are reviewed elsewhere.[6] These fragmentation processes include elimination of small neutral molecules, unzipping the polymer chain yielding monomer units or random chain cleavage. For example, initial decomposition of poly (vinyl chloride) is dominated by loss of HCl. Vinyl polymers such as polystyrene and poly (methyl methacrylate) decompose by unzipping the polymer chain to give high yields of respective monomers. Condensation polymers such as nylon will decompose by a combination of these three mechanisms.

The mechanism of thermal degradation depends on the polymer type, but the nature and relative yields of pyrolysis products are also related to pyrolyzer design and pyrolysis conditions. Critical factors affecting reproducible yield of characteristic pyrolysis fragments include the final pyrolysis temperature, the temperature rise time or rate of heating to this final temperature and sample size. With low final pyrolysis temperatures (300 to 600°C), fragments are generally monomers or other short-chain molecules (oligomers) related to the original polymer structure. As the final pyrolysis temperature increases, yields of these low molecular weight fragments may decrease because of fragmentation to even simpler molecules such as methane, carbon dioxide, and water. Photodiodes and low mass thermocouples are used to accurately measure this final pyrolysis temperature of filament or Curie point pyrolyzers[7].

The temperature rise time (TRT) is defined as the time interval required to heat the sample from ambient or its initial temperature to the final pyrolysis temperature. If a portion of the polymer sample is pyrolyzed during this heating interval, the duration of heating or TRT has a pronounced effect on the composition and yields of products in the pyrolyzate. In general, rapid rise times are preferred for maximum reproducibility of product yields and minimal side reaction of pyrolysis fragments.

In conjunction with temperature rise time, sample size considerations are important in determining the composition and yield of polymer fragments. Side reactions are more likely to occur with large samples which cannot be uniformly heated. Generally, microgram quantities are reproducibly pyrolyzed with rise times of 0.3 to 2 sec whereas larger samples require heating times of 5 to 30 sec.

Pyrolyzer Types

Temperature rise time and sample size are frequently dictated by pyrolyzer design. P-GC systems should be designed to minimize dead volume and cold spots while permitting rapid transfer of pyrolysis products to the GC column. Pyrolyzers in use for solids and high-boiling liquids are classified as either pulse-mode or continuous-mode. Curie point and filament pyrolyzers are in the first category and furnace units are operated in the continuous mode. Comparisons of various pyrolyzer types for study of polymer composition are presented in the literature.[8,9] Table 1 compares critical features of the above mentioned pyrolyzers as well as less common techniques (i.e., laser pyrolysis).

Curie Point

A ferromagnetic wire is heated to a specific temperature (Curie point) using a high-frequency induction coil. The Curie point is defined as the temperature at which the wire alloy becomes paramagnetic and its energy intake decreases. Commercially available pyrolyzers come with wires of alloy compositions to give Curie point temperatures of 300 to 1000°C.[10] A typical Curie point pyrolyzer cross-section is shown in Figure 1. Meuzelaar[11] has modified this design to give an automated Curie point pyrolyzer compatible with capillary chromatographic columns.

Factors affecting the TRT and final temperature have been investigated for Curie point pyrolyzers.[7,12] A major problem affecting reproducibility of Curie point pyrograms appears to be lack of quality control for wire composition resulting in nonreproducible final temperatures. By design, use of Curie point wires prohibits stepped-temperature pyrolysis, i.e., subjecting a sample to a series of increasing temperatures.

Filament or Ribbon

Nichrome and platinum wires in a variety of configurations have been used as pyrolyzer filaments. These filaments can be heated to a final pyrolysis temperature by application of a constant voltage. Filaments heated in this manner have slow, nonreproducible rise times (5 to 25 sec) resulting in poor pyrogram reproducibility. Capacitive boosted filaments are used to give faster rise times (7 to 10 msec) and improved repeatability.[7]

Soluble polymer samples are readily dip-coated from a volatile solvent to ensure uniform sample thickness on the filament. For insoluble polymer samples, coiled filaments are used with and without inserts such as quartz tubes or boats. This technique also increases the TRT because of the mass of the sample holder. Sample size and thickness affect pyrolysis results as previously discussed and these effects have been investigated for filament pyrolyzers.[13]

Ribbon pyrolyzers have more surface area than wire filaments to permit use of larger sample sizes. A commercially available ribbon probe (Chemical Data Systems) operates with a Wheatstone bridge circuit which functions to control the final pyrolysis temperature and TRT. This unit gives typical TRT of 10 msec to 600°C for microgram samples applied as thin films from solution. A V-groove ribbon is available to hold solid, insoluble polymer samples.

Furnace

Continuous mode furnace pyrolyzers consist of a pyrolysis tube or chamber, a sam-

Table 1
COMPARISON OF PYROLIZERS[8]

Characteristic Method	Sample size	Sample form requirements and limitations	Temperature range	Pyrolysis element temperature control	Catalytic reactions within pyrolysis chamber	Temperature rise time to 600°C	Repeatability	Disadvantages	Advantages
Boat	1—5000 μg	All forms 1. Liquids 2. Solids	Up to 1500°C	Continuously variable	Slight	20—50 sec	Fair	1. Slow rise time 2. Requiring large samples 3. Does not lend itself to capillary columns	1. Any material, liquid or solid 2. Widest temperature range 3. Wide sample size range 4. Can be purchased
Conventional filament and ribbon	10—1000 μg	1. Soluble in a solvent	Up to 1200°C	Continuously variable	Yes	5—25 sec	Fair	1. Slow rise time 2. Aging of metal 3. Solid material hard to apply	1. Most widely used 2. Wide temperature range 3. Purchase or built 4. Simple to apply sample
Capacitive boosted filament	5—10 μg	1. Soluble in a solvent	Up to 900°C	Continuously variable	No	7—10 msec	Good	1. Solid material must be soluble 2. Must use small sample size 3. Cannot be purchased	1. Fast rise time 2. Wide temperature range 3. Good repeatability
Curie point low power (30—100 W)	10—50 μg	1. Soluble in a solvent 2. Some crystals and powders	Up to 980°C	Limited to 356, 480, 510, 600, 770, 980°C	Yes	1.5 sec	Good	1. Solid material hard to apply 2. Pyrolysis chamber must be heated 3. Slow rise time	1. Accurate temperature 2. Good repeatability 3. Can be purchased
Curie point high power (> 1000 W)	10—50 μg	1. Soluble in a solvent 2. Some crystals and powders	Up to 980°C	Limited to 356, 480, 510, 600, 770, 980°C	Yes	70 msec		1. Solid material hard to apply 2. Pyrolysis chamber must be heated	1. Fast rise time 2. Accurate temperature 3. Good repeatability 4. Can be purchased

Table 1 (continued)
COMPARISON OF PYROLIZERS[8]

Characteristic Method	Sample size	Sample form requirements and limitations	Temperature range	Pyrolysis element temperature control	Catalytic reactions within pyrolysis chamber	Temperature rise time to 600°C	Repeatability	Disadvantages	Advantages
Vapor phase	0.001—10 μg	1. Volatile materials only	Up to 800°C	Continuously	No	N/A	Good	1. Only volatile materials	1. Good repeatability 2. Wide temperature range 3. Can be correlated with theory
Laser	500 μg or larger	1. Dark sample or media	Estimated 10°K	Not controlled	Slight	Estimated 10 μsec	Fair	1. Must use colored agent to pyrolyze 2. Complicated and expensive 3. Large sample size 4. Equilibrium temperature higher than other pyrolyzers 5. Laboratory hazards 6. Cannot be purchased 7. Complex fragmentation process	1. Pyrograms may be simple 2. Superior rise time 3. Not necessarily thermal degradation
Ribbon probe	0.1—5 μg	1. Soluble in a solvent 2. Some powders	Up to 1400°C	Continuously variable	Slight	10 msec	Good	1. Solid materials fall off 2. Must use small sample size	1. Excellent rise time 2. Wide temperature range 3. Can be purchased 4. Good repeatability

Reprinted from Table 2, *Chromatographia*, 5, 550, 1972. With permission of J. Q. Walker.

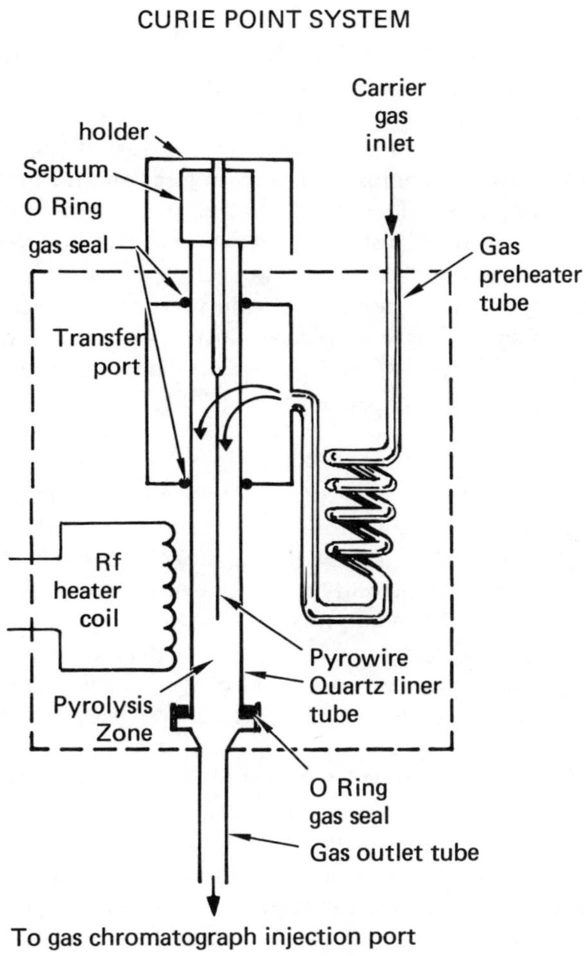

FIGURE 1. Schematic diagram of a typical Curie Point pyrolyzer.

ple holder, and a sample delivery system. Several designs are in use, so diagramming a "typical" pyrolyzer is impractical. The polymer, in a sample boat, is moved from a cool zone into a preheated pyrolysis zone swept by carrier gas to remove the decomposition products. The pyrolysis chamber must be composed of a noncatalytic surface to prevent interaction with the sample. A gas chromatographic injection port is the simplest type of furnace pyrolyzer. Gravity feed and magnetic push-rod assemblies are used for sample delivery.[4]

Previously discussed TRT and sample size effects are magnified for the furnace pyrolyzer. Sample sizes are in the milligram range. Sample boats and samples themselves exhibit finite temperature rise time after placement in the furnace, so the TRT for this type of pyrolyzer is in the 10 to 40-sec range. Rise times of this duration with large samples increase the probability of side reactions or recombination of pyrolysis products. From a chromatographic viewpoint, the large dead volume characteristics of furnace pyrolyzers causes peak broadening and limits the use of high resolution columns. On the positive side, however, the large sample size associated with furnace pyrolyzers permits accurate measurement of tars and nonvolatile pyrolysis products.

Other Pyrolyzers

Included in this category are laser and electric arc pyrolyzers. Laser rise times are on the order of μsec and final pyrolysis temperatures of 900 to 1200°C are common. These high temperatures frequently result in simpler fragmentation patterns than those obtained from other pyrolysis techniques. Fanter, Levy, and Wolf[14] discuss variables affecting laser pyrograms and compare data from laser and normal thermal degradation of polystyrene and polyethylene. Vanderborgh presents a thorough discussion of the laser pyrolysis process and sampling problems[15] and factors affecting reproducibility of laser induced pyrolysis.[16]

Electric discharge pyrolyzers, used in the early years of pyrolysis-gas chromatography, are not as prevalent as the commercial Curie point and filament units. A low current, high voltage pyrolyzer was used to study polymers and high boiling liquids.[17] Comparison with thermal pyrolyzer data is presented.

Correlation Trials

Effects of TRT, sample size, pyrolyzer configuration, and final pyrolysis temperature cause difficulties with interlaboratory comparison of pyrolysis data. International correlation trials were organized to evaluate the magnitude of the problem and determine the interlaboratory reproducibility available with today's pyrolysis instrumentation. In several publications,[18-20] the Pyrolysis Sub-Group of the Chromatography Discussion Group (Institute of Petroleum) reports conclusions of a series of European correlation trials. This work demonstrated that if pyrolysis and chromatographic conditions are specified, reproducible and identifiable pyrograms are obtained from one laboratory to another.

More recently, ASTM Committee E-19 sponsored three correlation trials to standardize interlaboratory P-GC data. This study resulted in a dual calibration procedure to achieve interlaboratory reproducibility.[21] A standard chromatographic column (e.g., Porapak Q®) is calibrated using a standard mixture. Carrier gas flow is adjusted until mixture components elute at prescribed retention times with a defined temperature. A standard polymer (e.g. styrene-isoprene copolymer) is pyrolyzed with adjustment of TRT and pyrolysis temperature to obtain a specified ratio of pyrolysis products. P-GC instrumentation calibrated in this manner should yield pyrograms which can be reproduced by another laboratory.

Qualitative Analysis

As in many other techniques, qualitative identification in P-GC involves comparison of unknown and standard pyrograms. The ease of identification is directly related to the simplicity and uniqueness of the fragmentation pattern or "fingerprint". For many polymer systems, only two or three characteristic peaks are needed for identification. Alekseeva[22] reports on the use of individual pyrolysis products for polymer identification. Examples presented include identification of polyisoprene, polychloroprene and butyl rubber in blends. Peak ratio techniques are useful to identify different brands of nitrile rubber or polybutadienes with minor differences in microstructure.

For complex molecules such as microorganisms, however, up to 15 pyrolysis products must be compared to identify an unknown. Computer programs are necessary for comparison of these complex pyrograms. Retention time matching techniques[23] and calculation of similarity coefficients[24] have proven useful. Pattern recognition techniques such as the linear learning method[25] have also been investigated for identification of polymer pyrograms.

Quantitative Analysis

For good quantitative analysis, a specific, well resolved pyrolysis fragment characteristic of the component to be quantified is necessary. Quantitation of polymer composition is often achieved by simply measuring the ratio of specific product peaks in the pyrogram. For acrylonitrile-butadiene-styrene terpolymer (ABS), correlation constants (K) are determined from pyrolysis of standards containing known concentrations of monomer units:

$$K_1 = \frac{P_s/P_a}{A_s/A_a} \qquad K_2 = \frac{P_b/P_a}{A_b/A_a} \qquad (1)$$

where P_s, P_a, and P_b are weight % styrene, acrylonitrile and butadiene monomer units in the terpolymer standard. Values A_s, A_a and A_b are monomer peak areas calculated from the standard pyrogram. Once K values are obtained, these same equations and relationship (2) are used to calculate the composition of any unknown ABS terpolymer:

$$P_a + P_b + P_s = 100 \qquad (2)$$

Selection of pyrolysis product peaks to be used for quantitative calculations is a problem particularly for samples giving complex fragmentation patterns. Computer techniques are reported[26] for selection and optimization of peaks to ensure high sensitivity and accuracy. Using a styrene-butadiene-2-vinylpyridene terpolymer, the same workers demonstrated use of external standardization to quantify polymer composition from P-GC data. The external standard was introduced into the chromatographic injection port prior to pyrolysis of the terpolymer.

Use of internal standards for quantitation in P-GC depends upon careful choice of the standard polymer. To be useful as an internal standard, a polymer must fragment quantitatively to a single peak when pyrolyzed. As in any gas chromatographic analysis, this major internal standard pyrolysis fragment must occur at a retention time without interference from sample pyrolysis products. This criterion is difficult to achieve for many polymer systems which give complex pyrograms. Poly (ethyl methacrylate) was used as an internal standard for analysis of acrylic resins and lacquers.[27]

Applications

Primary applications of P-GC involve synthetic polymers and biopolymers. Although too numerous to reference, major synthetic polymer applications include fingerprint identifications, quantitation of co-polymer or blend compositions, and structure determinations involving randomness, sequencing, and branching. The broad classification of biopolymers includes P-GC studies of bacteria, fungi, carbohydrates, proteins, sugars, and proteins. Early literature references to these applications are summarized in a bibliography published by Chemical Data Systems, Inc.[28] A more current summary of of P-GC applications is included in the recent overview of the technique by Irwin.[29] This overview also refers to forensic applications of P-GC and studies of fossil fuels, soil and extraterrestrial samples. In this area, references are available on P-GC applications to lignite[30] and oilshale.[31]

Instrumentation

Many pyrolyzers are designed and laboratory constructed to fill a specific need. Literature data presented in the tables of Section IV.2 were often obtained using commercially available pyrolyzers and gas chromatographs. Popular furnace pyrolyzers include Shimadzu, Hitachi, and Hamilton models; whereas, the Chemical Data Systems Pyroprobe is a convenient filament or ribbon system. Curie Point pyrolyzers referenced in the literature include models marketed by Shimadzu, Fischer, and Phillips or Pye.

REFERENCES

1. **Sherma, J.,** Gas chromatography principles and techniques, in *Handbook Series in Chromatography,* Vol. II, Zweig, G. and Sherma, J., Eds., CRC Press, Boca Raton, Fla., 1972, 7.
2. **Irwin, W. J.,** *J. Anal. Appl. Pyrolysis,* 1, 3, 1979.
3. **Levy, R. L.,** *Chromatogr. Rev.,* 8, 49, 1966.
4. **May, R. W., Pearson, E. F., and Scothern, D.,** *Pyrolysis — Gas Chromatography,* The Chemical Society, London, 1977.
5. **Jones, C. E. R. and Cramers, C. A., Eds.,** *Analytical Pyrolysis,* Elsevier, Amsterdam, 1977.
6. **Bamford, C. H. and Tipper, C. F. H., Eds.,** *Comprehensive Reaction Kinetics,* Vol. 14, Elsevier, Amsterdam, 1975.
7. **Levy, R. L., Fanter, D. L., and Wolf, C. J.,** *Anal. Chem.,* 44, 38, 1972.
8. **Walker, J. Q.,** *Chromatographia,* 5, 547, 1972.
9. **Alekeeva, K. V. and Khramova, L. P.,** *J. Chromatogr.,* 69, 65, 1972.
10. **Jackson, M. T. and Walker, J. Q.,** *Anal. Chem.,* 43, 74, 1971.
11. **Meuzelaar, H. C. L.,** *J. Chromatogr. Sci.,* 13, 12, 1975.
12. **Bühler, C. L. and Simon, W.,** *J. Chromatogr. Sci.,* 8, 323, 1970.
13. **Daniel, J. C. and Miekel, J. M.,** *J. Chromatogr. Sci.,* 5, 437, 1967.
14. **Fanter, D. L., Levy, R. L., and Wolf, C. J.,** *Anal. Chem.,* 44, 43, 1972.
15. **Vanderborgh, N. E.,** Laser induced pyrolysis techniques, in *Analytical Pyrolysis,* Jones, C. E. R. and Cramers, C. A., Eds., Elsevier, Amsterdam, 1977, 235.
16. **Vanderborgh, N. E. and Ristau, W. T.,** *J. Chromatogr. Sci.,* 11, 535, 1973.
17. **Sternberg, J. C. and Litle, R. L.,** *Anal. Chem.,* 38, 321, 1966.
18. **Coupe, N. B., Jones, C. E. R., and Perry, S. G.,** *J. Chromatogr.,* 47, 291, 1970.
19. **Coupe, N. B., Jones, C. E. R., and Stockwell, P. B.,** *Chromatographia,* 6, 483, 1973.
20. **Gough, T. A. and Jones, C. E. R.,** *Chromatographia,* 8, 696, 1975.
21. **Walker, J. Q.,** *J. Chromatogr. Sci.,* 15, 267, 1977.
22. **Alekseeva, K. V.,** *J. Anal. Appl. Pyrolysis,* 2, 19, 1980.
23. **Needleman, M. and Stuchbery, P.,** Identification of microorganisms by pyrolysis gas-liquid chromatography, in *Analytical Pyrolysis,* Jones, C. E. R. and Cramers, C. A., Eds., Elsevier, Amsterdam, 1977, 77.
24. **Stack, M. V., Donoghue, H. D., Tyler, J. E., and Marshall, M.,** Comparison of oral streptococci by pyrolysis GLC, in *Analytical Pyrolysis,* Jones, C. E. R. and Cramers, C. A., Eds., Elsevier, Amsterdam, 1977, 57.
25. **Küllik, E., Kaljurand, M., and Koel, M.,** *J. Chromatogr.,* 112, 297, 1975.
26. **Alishoyev, V. P., Berezkin, V. G., and Viktorovo, E. N.,** *J. Chromatogr.,* 108, 275, 1975.
27. **Esposito, G. G.,** *Anal. Chem.,* 36, 2183, 1964.
28. **Levy, E.,** A categorized bibliography on solids pyrolysis with selected references to vapor phase pyrolysis, in *CDS Technical Publication* 020175, Applications Laboratory Chemical Data Systems, Inc., Oxford, Pa.
29. **Irwin, W. J.,** *J. Anal. Appl. Pyrolysis,* 1, 89, 1979.
30. **Giam, C. S., Goodwin, T. E., Giam, P. Y., Rion, K. F., and Smith, S. G.,** *Anal. Chem.,* 49, 1540, 1977.
31. **Klesment, I.,** *J. Anal. Appl. Pyrolysis,* 2, 63, 1980.

I.3 LIQUID COLUMN CHROMATOGRAPHY

Since the *Handbook of Chromatography,* Volume II was published in 1972, great advances have taken place in the field of high performance liquid chromatography. These include improvements in both column packing materials and instrumentation. This presentation is designed to describe these advances without going into an in-depth description of the fundamentals of liquid chromatography. There are many fine books that have come out in recent years covering this subject.[1-6]

Publications covered in this section are those that pertain to conventional liquid chromatography as opposed to gel permeation chromatography. The latter subject is presented in another part of this book.

Application of liquid chromatography to the polymer field has generally been in the separation of polymer additives, monomers, and low molecular weight oligomers. If the polymers are in the molecular weight range of up to about 2000, these would be amenable to liquid chromatography.

Above that molecular weight, separation is best achieved by conventional gel permeation chromatography (GPC).

Column Packings

The late 1960s and early 1970s saw the introduction of the pellicular liquid chromatographic packings. These are spherical nonporous particles of nominally 37μ in diameter that are coated, or chemically bonded, with a thin film of the separating medium. Efficiencies are much improved over conventional adsorbents and ion exchangers. This is due to the rapid mass transfer of the solute into and out of the thin skin of the packing. However, because the separating medium is very thin, sample capacity is limited. Therefore, columns are often overloaded in order to obtain sufficient sensitivity of detection at the expense of resolution.

Following the pellicular packings, the totally porous silica particles were introduced. These are typically 5 and 10 μ in diameter. They offer high surface area which yield greater sample capacity. The small particle size results in increased efficiency. However, with the smaller particle size, back pressure is increased.

Various pendant groups, i.e., ion exchange, C_{18} and polar groups such as amino and nitrile can be attached to the silica particles to allow separation by the different modes of chromatography. The bulk of the separations currently carried out use this type of column packing material. Figure 2 shows a typical separation on a microparticulate reversed-phase high performance liquid chromatographic packing.

Detectors

Ultraviolet Absorption

The majority of liquid chromatographic detection is still being made by ultraviolet detectors. Manufacturers have made improvements in their units, but probably the most significant change has been the increased use of low-wavelength (< 220 nm) detection. Also fixed low-wavelength detectors are presently on the market.

Compounds not normally thought of as ultraviolet absorbers, such as aliphatic acids, amides, and halocarbons can be detected. Regular aromatic compounds generally have increased sensitivity of detection at these low wavelengths as compared with the more conventional 254 and 280 nm detector.

Refractive Index

The refractive index detector is based on the presence of components in the eluent causing a difference in the refractive index when compared against eluent alone. Of

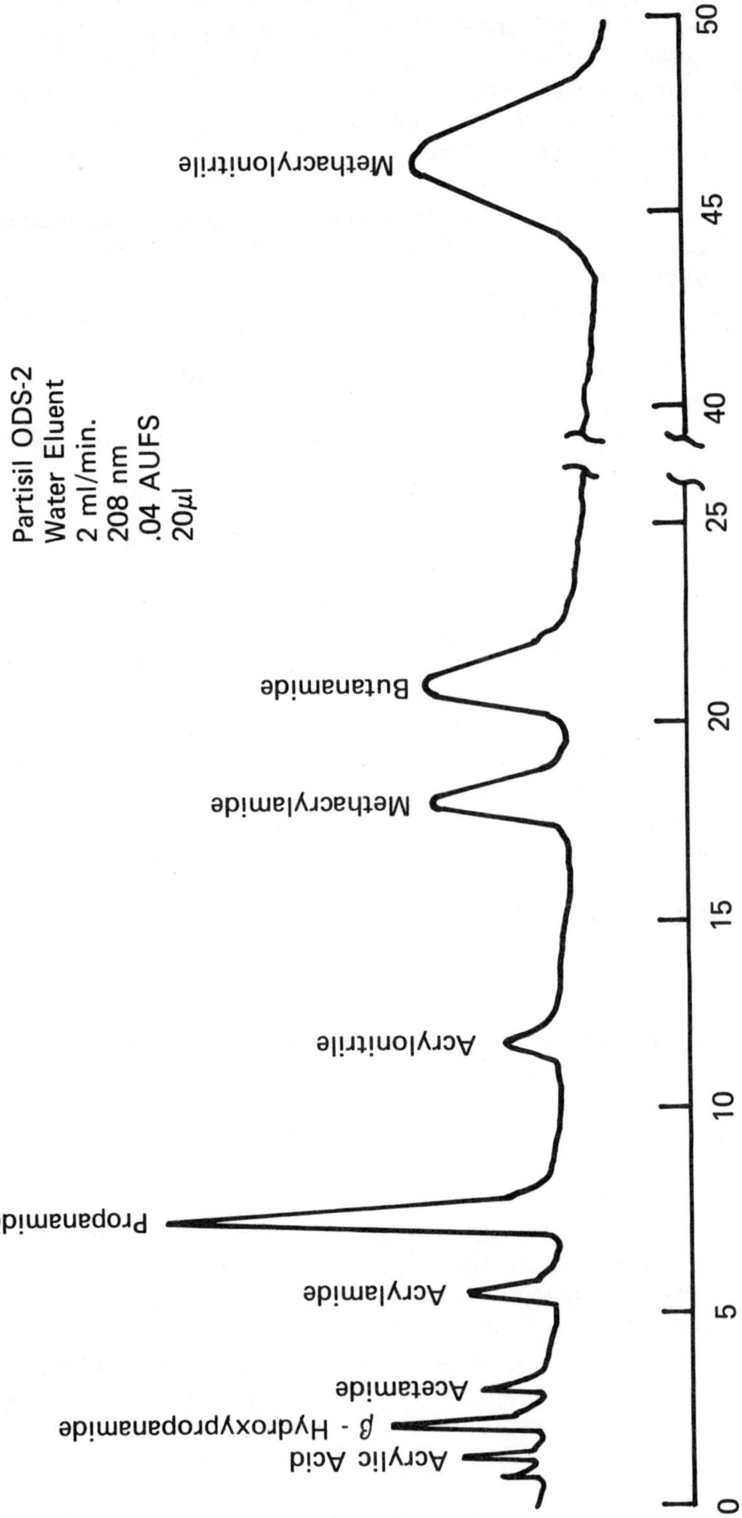

FIGURE 2. High-performance liquid chromatographic separation of acrylamide impurities on a microparticulate reversed-phase column.

all the detectors available, this comes the closest to being a universal detector. It is not as sensitive as most other detectors. Because the eluent plus component must be compensated for in the reference cell, the use of gradients is not practical.

Electrochemical

The last few years have seen the development of the electrochemical detector. Electrodes are encased in a low dead-volume cell and the sample flowing through the cell is allowed to oxidize or reduce at the electrode surface under controlled potential conditions. The electron current is monitored as a function of time to give a signal. This is then displayed on a recorder.

Since not all compounds are electroactive, it is best to use this detector in series with another, such as an ultraviolet detector, to be certain that some components do not pass through the cell undetected.

Most of the detectors presently available have been for the oxidative mode. More recently units for the reductive mode have come on the market.

Fluorescence

Certain chemicals will absorb light at one wavelength and emit it at a lower wavelength. This property is called fluorescence. If the chemical does not possess inherent fluorescence, this can be induced by derivatization with a fluorescent chromophore. Fluorescence detectors have the advantage of excellent selectivity and high sensitivity.

Infrared

Infrared detectors have been used with some success in liquid chromatographic applications. Since the eluent must be transparent at the wavelength at which the components are monitored, application of this detector is limited.

The two modes of liquid chromatography generally compatible with infrared detectors are absorption and normal-phase partition chromatography. Ion exchange and reversed-phase chromatographic separations employ aqueous-based eluents and therefore would dissolve infrared cell windows.

Injection Valves

With the particle size for column packings getting smaller and smaller, back pressure has increased. This has made sample introduction through a septum injection value more difficult. Problems with leakage, short septum life and injection under pressure are endemic.

To circumvent these problems most sample introduction is presently being done with loop injectors. With eluent continuously flowing through one side of the valve, sample is introduced at the injection port with the overflow going to waste. When the valve handle is turned, the slug of sample in the loop is placed into the eluent stream and thusly onto the column packing. Injectors are available which allow injection with a syringe of a lesser volume than the volume of the loop.

These injectors have generally been trouble-free with some rated as high as 10,000 psi. They may also be obtained with air-operators for use as automated injectors.

Pumps

Pumps for liquid chromatographic systems fall into two general classifications, constant pressure and constant displacement. Constant pressure pumps use a high-pressure gas, usually nitrogen to pressurize a cylinder containing the eluent. This pressure forces the eluent through the liquid chromatographic column. One of the disadvantages of this pumping system is that the gas becomes dissolved in the eluent. When the eluent reaches the detector which is near ambient pressure, the gas will expell from the eluent giving rise to bubbles in the detector cell.

This problem can be circumvented by using an intermediate fluid and a rigid piston. Also, by using a larger piston on the intermediate solvent relative to the piston for the eluent, the pressure to the column can be amplified without going to excessive pressures on the intermediate fluid.

It is more difficult to form gradients with these types of pumps. Also, back-pressure may change due to temperature fluctuations, solvent viscosity differences and particulates on the column. This will result in a corresponding change in flow rate. However, they are much less expensive than constant displacement pumps.

Constant Displacement

Constant displacement pumps get around many of the disadvantages of constant pressure pumps, but they are more expensive.

They fall into two main categories, the reciprocating pump and the syringe pump. In the former, solvent is pulled into the pumping chamber through a check valve. On the compression stroke, the inlet check closes and solvent is forced through the outlet check valve onto the column. The reciprocating pump is relatively inexpensive. Pressure limit controls and pulse dampeners will add to the cost. Because they continuously draw solvent into the pump, there is no volume limit they can pump. Their main disadvantages are as the piston seals wear, particulates from the seals may get into the check valves. With volatile solvents, these pumps have the tendency to cavitate. Because solvent delivery is cyclical, these pumps generally give a pulse. This can be overcome by the use of the pulse dampeners. However, these are undesirable when changing solvents or in the use of gradient elution. In some of the more expensive pumps the pulse can be reduced or eliminated by employing two pistons 180° out of phase.

Syringe

The syringe pump consists of a piston displacing the eluent in a cylinder. The flow rate is controlled by varying the rotation rate of the motor. Using two of these pumps gradients can be generated. These types of pumps give essentially no pulse.

REFERENCES

1. **Engelhardt, H.,** *High-Performance Liquid Chromatography: Chemical Laboratory Practice,* Springer-Verlag, Berlin, 1979.
2. **Hamilton, R. J. and Sewell, P. A.,** *Introduction to High Performance Liquid Chromatography,* Wiley-Interscience, New York, 1978.
3. **Huber, J. F. K.,** *Instrumentation for High-Performance Liquid Chromatography,* Elsevier, Amsterdam, 1978.
4. **Knox, J. H.,** *High-Performance Liquid Chromatography,* Edinburgh University Press, 1978.
5. **Pryde, A. and Gilbert, M. T.,** *Application of High-Performance Liquid Chromatography,* Chapman-Hall, London, 1978.
6. **Snyder, L. R. and Kirkland, J. J.,** *Introduction to Modern Liquid Chromatography,* 2nd ed., Wiley-Interscience, New York, 1979.

I.4 SIZE EXCLUSION CHROMATOGRAPHY

Since the introduction of the first commercial standard gel permeation chromatograph (GPC) by Waters Associates in 1965, the technique has become the prime analytical method to determine the molecular weight of polymers. Recently, GPC took the advantage of the revolutionary technical advances and theory developed in the early 1970s and evolved into a whole new generation of instrumentation. The analysis time by the standard GPC takes hours, while the new high speed GPC does it in minutes.

GPC is also known by many other names. Historically, exclusion chromatography is called gel filtration chromatography (GFC) with the aqueous media, and is called GPC when organic solvents are involved. The American Society for Testing and Materials has adopted the name of size exclusion chromatography (SEC) which will be used in this chapter.

This chapter is concerned with the high speed SEC in organic solvents. The technique of GFC has been well covered by Pharmacia Fine Chemicals® published literatures and its annual "Literature References".

Theory and Mechanism

SEC is one of the many forms of liquid chromatography (LC). It separates molecules by their hydrodynamic volume in solution.

A bed of SEC column is composed of porous material of various sizes. When molecules of very large size are introduced into the column, they will not diffuse into the pores to any great extent and elute earlier. Smaller molecules have to dffuse in and out of the pores to a greater degree and elute later.

Diffusion of solutes into the pores is sufficiently faster than elution rate and dynamic diffusion equilibrium can be assumed. The separation may be described as

$$V_r = V_o + KV_i \qquad (3)$$

Where V_r: Retention or elution volume; V_o: Void volume; V_i: Pore volume; and K: Distribution coefficient.

The value of K is between 0 and 1. K = 0 when a solute is totally excluded from the pores, and K = 1 when a solute totally permeates the pores. This limiting values of K provides a unique property to the SEC where the total elution volume for all components is fixed for any particular column system.

Instrumentation

SEC instrumentation in general consists of a pump, injector, column, detector, and some sort of data collecting and processing system. Since pump and injector are the same as any other LC, they are not discussed here. However, column and its selection and some detectors which are specific to SEC are covered in the following discussion.

Column Selection and Packing

Since the separation in SEC is based upon distributions of molecular size of the solute and the pore size of the packing, the match of the average pore size to the size range of the solute is of vital importance for optimum separation. A semi-log plot of molecular size vs. elution volume results in a calibration curve as shown in Figure 3.

The slope of the calibration curve depends on the distribution of the pore. The narrower the distribution, the flatter is the slope, and the better the resolution, however,

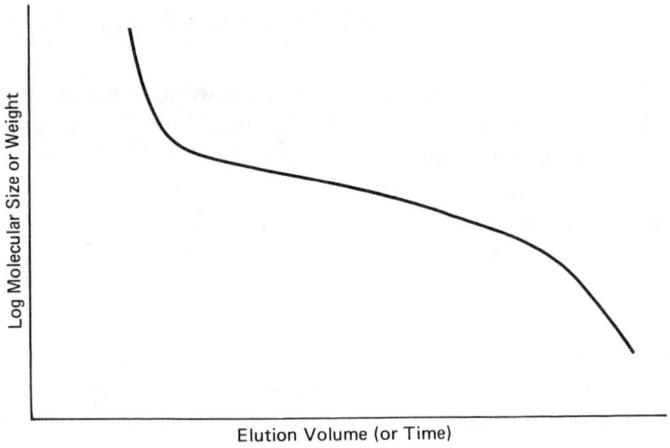

FIGURE 3. Diagram of a typical SEC calibration curve.

the operating molecular size range is smaller. Connection of several columns of different pore size packings is necessary to cover a broad molecular size range of polymeric samples.

The most commonly used column packings for SEC are polystyrene cross-linked with divinylbenzene and porous glass or silica with or without surface modification. The two types of packings can be characterized as follows:

Characteristic	Cross-linked polystyrene	Porous glass or silica
Mechanical stability	Poor	Good
Resolution	Good, especially for low molecular weight compounds	Very poor for molecular weight below 5000
Adsorption	Generally not a problem	Always a problem
Availability	In prepacked columns only	Packings are commercially available
Maintenance	Should not be dried out	Less care
Temperature limit	150°C	Not a concern for all practical purposes
Solvent change	Not recommended	Easy
Solvent	Organic Only	Organic or aqueous

Commercially available packings or prepacked columns are listed in Table 2 of Section II.1.

Detectors

A detector is usually connected to the outlet of a column to monitor either the concentration or composition of the effluent. The detectors commonly used in LC include differential refractometers, UV, and infrared detectors. Other special detectors such as fluorescence, electrochemical, flame ionization, conductivity, mass spectrometric and colorimetric detectors have found increasing applications. However, only a couple of detectors which are specifically applicable to SEC are discussed here.

Viscometric Detectors

A universal calibration scheme was proposed by Grubisic, Rempp, and Benoit in 1967.[1] It was obtained by plotting 1n ([η]M) against elution volume or time where [η] is the intrinsic viscosity and (M) the molecular weight. Such a relation holds for many polymers regardless of their chemical structures and compositions. The application of this universal calibration requires the measurements of the concentration as well as the intrinsic viscosity of the fractions as they elute from the SEC column. There are two devices that can be used to measure the viscosity of the eluent. One is a Ubbehlode type viscometer which collects the fractions and measures the efflux times of each fraction. It is noncontinuous and does not measure viscosity point by point. Another type of viscometric detector was developed by Ouano.[2] It measures the pressure drop across a capillary by a pressure transducer. Since pressure drop of a fluid flowing through a capillary is proportional to the viscosity for a given system and flow rate, this device serves as a truly flow-thru detector to measure the viscosity of the eluting fractions continuously.

Light Scattering Detector

The SEC is most commonly used to determine molecular weight (MW) and molecular weight distribution (MWD) of polymers. However, tedious column calibration is required for those measurements. It would be ideal if a detector can determine the MW of each fraction as it elutes from the column. It is possible to do that with the introduction of a low angle laser light scattering photometer, originally developed by Kaye[3] and now commercially available from Chromatix.

The weight average molecular weight (MW) can be calculated from the fluctuation theory of light scattering as follows:

$$\frac{KC}{R_o} = \frac{1}{M_W} + 2A_2C \tag{4}$$

$$K = \frac{2\pi^2 n_o^2}{\lambda^4 N}\left(\frac{dn}{dc}\right)^2 (1 + \cos^2\theta) \tag{5}$$

$$R_o = \left\{\left(\frac{I_s}{I_o}\right)_{sample} - \left(\frac{I_s}{I_o}\right)_{solvent}\right\} D\,(\sigma' 1')^{-1} \tag{6}$$

Where C: Solute concentration in g/mℓ
n_o: Refractive index of the solvent
$\frac{dn}{dc}$: Specific refractive index increment (refractive index of solution as a function of solute concentration)
λ: Wavelength of incident light
N: Avagadros' number
A_2: Second virial coefficient D, o′, 1′: Photometer constants
Θ: Scattering angle
I_s: Scattering light intensity
I_o: Incident light intensity

Since all the terms except $(I_s)_{sample}$ can be premeasured, the light scattering photometer can determine the MW of each fraction by measuring the $(I_s)_{sample}$ as the sample fraction elutes from the column. MWD of a polymer can then be constructed from the MW of each individual fraction.

Calibration and Data Treatment

SEC is primarily used for the determination of MW and MWD of polymers. The column used for this application has to be calibrated. A series of MW standards of narrow distribution is first run on the column. A calibration curve can then be constructed by plotting the logarithm of the MW of the standards versus their elution volumes. The MW and MWD of a sample can be calculated from its chromatogram using the calibration curve.

The availability of the type of narrow distribution standards is rather limited. Since different types of polymer of the same MW may have different molecular size, the MW and MWD of a polymer calculated from a calibration curve established with another type of polymer can lead to erroneous results. In order to avoid the necessity for narrow distribution standards and also to facilitate the calibration, broad distribution polymers have been used to prepare calibration curves.[4-6]

When monodispersed compounds elute from a column, zone-spreading or peak broadening always occurs due to multiple paths of the column, molecular diffusion and mass transfer between the stationary and mobile phases. Correction of the zone-spreading in other LC separations is not important since accurate quantitation can still be made as long as the peaks are well resolved. However, a polymer sample consists of thousands of molecular species of various sizes, and its chromatogram is a composite of overlapping peaks of all of those molecular species which cannot be resolved by the column. Both ends of the peak due to zone-spreading can be interpreted as components which do not even exist. Inclusion of those nonexisting components in the calculation of MW and MWD may result in serious error. This broadening effect can be minimized by increasing the efficiency of the column but can never be totally eliminated. Another approach is to remove it mathematically. Tung and many others[7-11] have proposed various techniques to evaluate and derive correction methods.

Applications

The number of published literatures on SEC is enormous and exceeds more than a thousand during the period of 1970 to 1979. It is impractical to list all the applications and experimental conditions as done in chapters of other chromatographies of this book. It is also of little value since the types of column and solvent, variation of temperature, and flow rate, etc. used in SEC are very limited. The Fundamental Review issues[12-15] of the *Journal of Analytical Chemistry* and three issues of "Bibliography on Liquid Exclusion Chromatography (Gel Permeation Chromatography)"[16-18] published by the American Society for Testing and Materials provide excellent sources for those who are interested in a particular subject or application.

REFERENCES

1. Grubisic, Z., Rempp, P., and Benoit, H., *J. Polymer Sci.*, B5, 753, 1967.
2. Ouano, A. C., *J. Polymer Sci.*, A1, 10, 2169, 1972.
3. Kaye, W. and Havlik, A. J., *App. Opt.*, 12, 541, 1973.
4. Balke, S. T., Hamielec, A. E., Leclair, B. P., and Pearce, S. L., *Ind. Eng. Chem. Prod. Res. Dev.*, 8, 54, 1969.
5. Frank, F. C., Ward, I. M., and Williams, T., *J. Polymer Sci.*, A2, 6, 1357, 1968.
6. Weiss, A. R. and Cohn-Ginberg, E., *J. Polymer Sci.*, A2, 8, 148, 1970.
7. Tung, L. H., *J. Appl. Polymer Sci.*, 10, 375, 1271 (1966), 13, 775 (1969).
8. Pickett, H. E., Canton, M. J. R., and Johnson, J. F., *J. Polymer Sci.*, C21, 67, 1968.
9. Hamielec, A. E. and Ray, W. H., *J. Appl. Polymer Sci.*, 13, 1319, 1969.

10. **Dawidowicz, A. and Sokolowski, S.,** *J. Chromatogr.*, 125, 428, 1976.
11. **Vozka, S. and Kubin, M.,** *J. Chromatogr.*, 139, 225, 1977.
12. **Zweig, G. and Sherma, J.,** *Anal. Chem.*, 44(5), 42R, 1972.
13. **Zweig, G. and Sherma, J.,** *Anal. Chem.*, 46(5), 73R, 1974.
14. **Gaylor, V. F., James, H. L., and Weetall, H. H.,** *Anal. Chem.*, 48(5), 44R, 1976.
15. **Gaylor, V. F. and James, H. L.,** *Anal. Chem.*, 50(5), 29R, 1978.
16. **James, H. L. and Mitchell, D. M.,** Bibliography on Liquid Exclusion Chromatography, through 1972, AMD 40, American Society for Testing and Materials, Philadelphia, Pa.
17. **James, H. L. and Mitchell, D. M.,** Bibliography on Liquid Exclusion Chromatography, 1972-1975, AMD 40-51, American Society for Testing and Materials, Philadelphia, Pa.
18. **James, H. L.,** Bibliography on Liquid Exclusion Chromatography, 1976-1979, AMD 40-52, American Society for Testing and Materials, Philadelphia, Pa.

I.5 THIN-LAYER CHROMATOGRAPHY

With the increased use of high pressure (high performance) liquid chromatography (HPLC) over the past decade, thin-layer chromatography's growth has slackened although several books[1-6] and review articles[7-9] are available on this subject. The former technique has the advantage of offering a better means of quantitation. However, the use of densitometry for *in-situ* quantitation on the thin-layer plates has replaced visual examination for most quantitative applications.

In recent years high performance TLC (HPTLC) on chemically bonded layers has shown decided advantages. It offers both better sensitivity and efficiency, higher speed, and the use of less solvent. Also, plates are available for the reversed-phase mode (RPTLC). Although not giving a direct correlation with reversed-phase liquid chromatography it does give sufficient information as to what solvent systems would be appropriate to try. More importantly, it would indicate which component(s) are staying at the top of the HPLC column.

REFERENCES

1. **Kirchner, J. G.,** Thin-Layer Chromatography, in *Techniques of Chemistry,* Vol. 14, 2nd ed., Wiley-Interscience, New York, 1978, 1137.
2. **Smith, I. and Seakins, J. W. T., Eds.,** Paper and thin-layer chromatography, in *Chromatographic and Electrophoretic Techniques,* Vol. 1, 4th ed., Year Book, Medical Publications, Chicago, Ill., 1976.
3. **Mikes, O., Ed.,** *Laboratory Handbook of Chromatographic and Allied Methods,* Wiley, Chichester, England, 1978, 820.
4. **Touchstone, J. C., and Sherma, J.,** *Densitometry in Thin-Layer Chromatography: Practice and Applications,* Wiley, Chichester, England, 1979, 764.
5. **Goetz, W., Sachs, Al, and Wimmer, H.,** *Thin-Layer Chromatography,* Gustav Fischer Verl., Stuttgart, 1978, 120.
6. **Gasparic, J. and Churacek, J.,** *Laboratory Handbook of Paper and Thin-Layer Chromatography,* Wiley, Chichester, England, 1978, 362.
7. **Mott, O. and Novotny, L.,** *Laboratory Handbook of Chromatographic and Allied Methods,* Wiley, Chichester, England, 1979, 462.
8. **Prochazka, Z., Hejtmanek, M., Sebesta, K., and Tomasek, V.,** *Laboratory Handbook of Chromatographic and Allied Methods,* Wiley, Chichester, England, 1979, 64.
9. **Stahl, E.,** *J. Chromatogr.,* 165, 59, 1979.

Section II
Practical Applications

II.1 PRODUCTS AND SOURCES OF CHROMATOGRAPHIC MATERIALS

This Section is presented as a supplement to Section II, Volume II of the *Handbook Series in Chromatography.* Included is a listing of sources for chromatographic columns compiled from the "1980—1981 Laboratory Guide", *Anal. Chem.,* August, 1980.

Sources listed are not intended as editors' endorsement since other sources may be available.

Readers interested in complete listings of chromatographic instruments, manufacturers, and accessories are referred to the Laboratory Guide referenced above.

Table 1
PRODUCTS AND SOURCES OF CHROMATOGRAPHIC MATERIALS

	GC		LC			
Supplier	Column packings	Packed[a] columns	Column packings	Packed columns	TLC	SEC
Alltech Associates, Inc.®	x	x	x	x	x	
Altex Scientific, Inc.®	x	x		x		x
Amicon Corp.®				x		
Analtech, Inc.®				x	x	
The Anspec Co., Inc.®	x	x	x		x	x
Antek Instruments, Inc.®	x	x				
Applied Science Division®	x	x	x	x	x	
J. T. Baker Chemical Co.®	x				x	
Beckman Instruments, Inc.®	x	x		x		
Benson Co.®	x	x	x	x		
James G. Biddle Co.®	x					
Bioanalytical Systems, Inc.®		x		x		
BioRad Laboratories®	x	x	x	x	x	x
Brinkman Instrument, Inc.®		x			x	
Buck Scientific, Inc.®	x	x				
Cal-Glass for Research, Inc.®		x				
Chromatix, Inc.®		x		x		x
Chromatronix, Inc.®		x	x	x	x	
Chrompack Nederland BV®	x	x	x	x		
Coast Engineering Laboratory®	x					
Corning Glass Works®		x			x	
Dionex Corp.®	x	x				
E. I. duPont deNemours & Co. Inc.®	x	x	x	x		x
Carlo Erba Strumentazione®	x					
Fisher Scientific Co.®	x	x			x	
Floridin Co.®	x		x		x	
Fotodyne, Inc.®					x	
Foxboro Analytical®	x	x	x	x	x	
Galileo Electro-Optics Corp.®		x				
A. Gallenkamp and Co., Ltd.®		x				
Gelman Sciences, Inc.®						x
Glenco Scientific, Inc.®		x		x		
W. R. Grace & Co.®	x		x		x	
Hamilton Co.®	x		x	x		
Handy & Harman Tube Co.®		x				
HETP®				x		

Table 1 (continued)
PRODUCTS AND SOURCES OF CHROMATOGRAPHIC MATERIALS

	GC		LC			
Supplier	Column packings	Packed[a] columns	Column packings	Packed columns	TLC	SEC
Hewlett-Packard, Avondale Div.®	x	x		x		
Hi-Q Filter Products Co.®	x					
ICN Nutritional Biochemicals®	x	x	x	x	x	
Instruments SA, Inc.®	x	x				
Isolab, Inc.®		x				
ISCO-Chemresearch®				x		
Japan Spectroscopic Co., Ltd.®				x		
J&W Scientific, Inc.®		x				
Jasco, Inc.®	x	x		x		
Johns-Manville®	x		x	x		
Jones Chromatography, Inc.®	x	x	x	x		
Kimble Div.®		x				
Kontes®	x					x
Kratos, Inc.,®						
Schoeffel Instr. Div.				x		
Lab Crest Scientific Glass Co.®		x				
Lab Glass, Inc.®		x				
Laboratory Data Control®	x	x	x	x		
Lachat Chemicals, Inc.®	x	x		x		
L. C. Company, Inc.®	x	x	x	x		
Lion Technology, Inc.®	x	x	x	x	x	
LKB Instruments, Inc.®	x	x				
Mallinckrodt, Inc.®	x				x	
Mark Instrument Co.®	x	x				
MCB Manufacturing Chemists, Inc.®	x	x		x	x	
MCB Reagents®	x	x	x	x	x	
Micromeritics Instrument Corp.®	x	x	x	x		
Nalge Co.®				x		
Perkin-Elmer Corp.®	x	x		x		x
Pharmacia Fine Chemicals®	x	x			x	x
Phase Separations Ltd.®	x	x	x	x	x	
Pierce Chemical Co.®	x					x
Polysciences, Inc.®	x		x			
Prolabo®	x	x		x	x	
QuadRex Corp.®		x				
Rainin Instrument Co., Inc.®	x	x	x	x	x	
Regis Chemical®	x	x		x		
Reliance Glass Works, Inc.®				x		
RFR Corp.®	x	x	x			
Schleicher & Schuell, Inc.®	x				x	
The Separations Group (Vydac)®	x	x	x	x		
Shandon Southern Instruments, Inc.®	x	x				
Sigma Chemical Co.®			x		x	
Spectra Physics®	x	x	x	x		
Supelco, Inc.®	x	x		x	x	
Toyo Kagaku Sangyo Co., Ltd.®						x
Tridom Chemical, Inc.®						x
Tudor Scientific Glass Co.®		x				
Unimetrics Corp.®	x	x		x		

Table 1 (continued)
PRODUCTS AND SOURCES OF CHROMATOGRAPHIC MATERIALS

Supplier	GC Column packings	GC Packed[a] columns	LC Column packings	LC Packed columns	TLC	SEC
Union Carbide Corp.,® Linde Specialty Glass	x				x	
Universal Scientific, Inc.®	x	x	x	x	x	
Varian Instrument Group®	x	x	x	x		x
Waters Associates®			x	x		x
Whatman, Inc.,® Chemical Separation Div.	x	x	x	x	x	
Wheaton Scientific®		x				
Wilk Instruments®		x				
Yound Laboratories®		x				

[a] Includes capillary and open tubular (SCOT) columns.

Table 2
COMMERCIALLY AVAILABLE PACKINGS FOR HIGH SPEED STERIC EXCLUSION CHROMATOGRAPHY (GEL PERMEATION CHROMATOGRAPHY) WITH ORGANIC SOLVENTS

Crosslinked Polystyrene

Type	MW exclusion limit or range	Average pore size (Å)
μ-Styragel[a]	700	100
	$(0.05—1) \times 10^4$	500
	$(0.1—20) \times 10^4$	10^3
	$(1—20) \times 10^4$	10^4
	$(1—20) \times 10^5$	10^5
	$(5—>10) \times 10^5$	10^6
HSG[b]	4×10^2	
	4×10^3	
	4×10^4	
	2×10^5	
	8×10^5	
	2×10^6	
	8×10^6	
	10^7	
Shodex[c]	1×10^3	
	5×10^3	
	7×10^4	
	5×10^5	
	5×10^6	
	5×10^7	
TSK Gel[d]	$(0.05—1) \times 10^3$	40
	$(0.05—6) \times 10^3$	250
	$(0.1—60) \times 10^3$	1500
	$(0.5—40) \times 10^4$	10^4
	$(0.2—400) \times 10^4$	10^5

Table 2 (continued)
COMMERCIALLY AVAILABLE PACKINGS FOR HIGH SPEED STERIC EXCLUSION CHROMATOGRAPHY (GEL PERMEATION CHROMATOGRAPHY) WITH ORGANIC SOLVENTS

Crosslinked Polystyrene

Type	MW exclusion limit or range	Average pore size (Å)
	4×10^7	10^6
	4×10^8	10^7
	1.5×10^3—10^7	
	Silica	
Porasil[e]		
Type A	4×10^4	< 100
Type B	2×10^5	150
Type C	4×10^5	300
Type D	1×10^6	600
Type E	1.5×10^6	1200
Type F	$> 4 \times 10^6$	> 1500
Spherosil[f]		80
		140
		360
		680
	$> 4 \times 10^6$	1100
LiChrospher[g] S1	6×10^4	100
	5×10^5	500
	2×10^6	1000
	$> 3.5 \times 10^6$	4000
SEC[h]	200—5000	60
	1×10^5	100
	5×10^5	500
	1×10^6	1000
	$> 4 \times 10^6$	4000
EM Gel Type S1[i]	50,000	200
	400,000	500
	1,000,000	1000
	Glass	
Bio-Glass[j]		
Type 200	1×10^5	200
Type 500	4×10^5	500
Type 1000	1×10^6	1000
Type 1500	$> 4 \times 10^6$	1500
Type 2500	$> 4 \times 10^6$	2500
Controlled-Pore Glass[k]		
CPG 00075	3,000—30,000	75
000120	7,000—130,000	120
000170	12,000—400,000	170
000240	22,000—1,200,000	240
000350	40,000—4,000,000	350
000400	50,000—5,500,000	400
000500	70,000—10,000,000	500
000700	130,000—30,000,000	700
001000	250,000—100,000,000	1000

Table 2 (continued)
COMMERCIALLY AVAILABLE PACKINGS FOR HIGH SPEED STERIC EXCLUSION CHROMATOGRAPHY (GEL PERMEATION CHROMATOGRAPHY) WITH ORGANIC SOLVENTS

Crosslinked Polystyrene

Type	MW exclusion limit or range	Average pore size (Å)
001400	400,000—300,000,000	1400
001600	500,000—400,000,000	1600
002000	700,000—900,000,000	2000
003000	900,000—1,900,000,000	3000

[a] Waters Associates®
[b] Shimadzu (Japan)®
[c] Showa Denko KK® (Japan)
[d] Toyo Soda Co.® (Japan), also available from Varian Associates®
[e] Waters Associates®
[f] Rhone Poulenc® (France)
[g] E. Merck®
[h] DuPont®
[i] E. Merck®
[j] BioRad Laboratories®
[k] Corning (Pierce), Electro-Nucleonics, Inc.®

II.2. SUBJECT INDEX

Section III
List of Tables

III.1 GAS CHROMATOGRAPHY TABLES

III.2 PYROLYSIS-GAS CHROMATOGRAPHY TABLES

III.3 LIQUID COLUMN CHROMATOGRAPHY TABLES

III. 4 THIN-LAYER CHROMATOGRAPHY TABLES

Section IV
Tables for Chromatographic Data

IV. 1 GAS CHROMATOGRAPHY

INTRODUCTION

For the most part, tables are arranged in alphabetical order by polymer or class of compound. The reader is referred to the List of Tables in Section III or the master index for specific compounds.

Please note that n.a. = not available from the original source; T_r = retention time or temperature; and I_r = retention index.

Table GC 1
ACCELERATOR COMPARISON BY MODEL COMPOUND VULCANIZATION

The efficiency of each of 20 accelerators to convert sulfur into sulfidic products has been evaluated by Model Compound Vulcanization. The sulfidic products can be isolated and identified by gas-liquid chromatography. The sulfidic products can be separated into groups of mono-, di-, and trisulfides, with several isomeric products within each group. The analysis of monosulfidic product yields and the isomer distribution within the monosulfide group is sufficient to characterize an accelerator.

Column packing	15% DC-200
Temperature program (°C)	
Initial	140
Final	200
Rate (°C/min)	1
Gas; flow rate	Helium; 60
Column	
Length, m	3.66
Diameter, mm	32
Material	—

Lautenschlaeges, F. K. and Zeeman, P., *Rubber Chem. Technol.*, 22, 1030, 1979.

Table GC 2
AMINE ANTIDEGRADANTS IN RUBBER COMPOUNDS

Extract 15 to 20 g of sample in a Soxhlet extractor with 100 mℓ of acetone for 16 hr. Evaporate the extract to a total volume of 10 mℓ at 50°C under reduced pressure. Chromatograph a 20-μℓ aliquot.

Column packing	P1
Temperature (°C)	290
Gas; flow rate (mℓ/min)	He; 150
Column	
Length, m	2
Diameter, in.	1/4
Material	Al
Detector	n.a.

Compound	Retention time (min)	Relative retention time[a]
1,2-Dihydro-2,2,4-trimethyl-6-ethoxyquinoline	1.77	0.22
N,N′-Dimethyl-*N*,*N*-di(1-methylpropyl)-*p*-phenylenediamine	2.49	0.31
N-Isopropyl-N′-phenyl-*p*-phenylenediamine	4.83	0.60
Di-(1-methylpropyl)-*p*-phenylenediamine	1.73	0.20
N,N′-Di-3-(5-methylheptyl)-*p*-phenylenediamine	7.03	0.88
N-Phenyl-2-naphthylamine	8.03	1.00
N,N′-Di-2-octyl-*p*-phenylenediamine	11.87	1.48
N-Cyclohexyl-N′-phenyl-*p*-phenylenediamine	17.55	2.19
N,N′-Diphenyl-*p*-phenylenediamine	20.80	2.59

Note: Column packing — P1 = 20% Apiezon L® on Chromsorb W® (30/60 mesh).

[a] Relative to *N*-phenyl-2-naphthylamine.

Wise, R. W. and Sullivan, A. B., *Rubber Age*, 91, 773, 1962.

Table GC 3
OCCLUDED SOLVENTS IN ADHESIVES

Insert 10 mm × 275 mm adhesive strips into a vaporizing unit attached directly to the inlet port of a gas chromatograph. Heat the unit for 10 min at 70°C to vaporize the solvent. Inject the vaporized solvents directly onto the column over a 3-sec interval.

Column packing	P1	P2
Temperature, (°C)	57.5	85.5
Gas, flow rate (mℓ/min)	N_2; 30	N_2; 30
Column		
Length, m	1.5	1.5
Diameter, mm I.D.	4	4
Material	SS	SS
Detector	FI	FI

Compound	Retention time (min)	
n-Pentane	1.8	
2-Methylpentane	3.5	
3-Methylpentane	4.0	
n-Hexane	4.5	
Methylcyclopentane	6.0	
Benzene		3.2
Cyclohexane		3.8
Toluene		7.5

Note: Column packing — P1 = 15% Squalane® on Shimalite® (60/80 mesh).

Suzuki, M., Tsuge, S., and Takeuchi, T., *Anal. Chem.*, 42, 1705, 1970.

Table GC 4
RESIDUAL ACRYLAMIDE MONOMER IN POLYELECTROLYTES

Shake 1 g of powdered polyelectrolyte with 10 mℓ of 80:20 methanol-water for 24 hr. Chromatograph a 5-μℓ portion of the clear extract. The method is capable of detecting 0.0004% of acrylamide in the polymer.

Column	P1
Temperature (°C)	180
Gas: flow rate (mℓ/min)	N_2; 32
Column	
Length, m	1
Diameter, mm, I.D.	3
Material	Glass
Detector	FI

Compound	Retention time (min)
Acrylamide monomer	4.33

Note: Column packing — P1 = 20% Carbowax 20M® on Chromosorb W-AW-DMCS® (60/80 mesh).

Croll, B. T., *Analyst*, 96, 67, 1971.

Table GC 5
UNREACTED ALDEHYDES AND PHENOL IN AQUEOUS SOLUTIONS OF PHENOLIC RESINS

Column packing	P1	P2
Temperature (°C)	130	130
Gas; flow rate (mℓ/min)	He; 105	He; 36
Column		
Length, ft	12	16
Diameter, in.	1/4	1/4
Material	Cu	Cu
Detector	TC	TC

Compound	Retention time (min)[a]	
Formaldehyde		0.4
Ether	0.3	0.6
Methanol	0.1	1.5
Acrolein	0.2	1.9
Water	<0.1	2.1
1-Butanol (Int. Std. P2)	0.6	5.5
Furfural	1.8	
Phenol	3.6	
m-Cresol (Int. Std. P1)	6.3	

Note: Column packing — P1 = 10% GE silicone SF-96 on Fluoropak®;
P2 = 10% sucrose octaacetate on Teflon 6®.

[a] Relative time from air peak.

Stevens, M. P., *Anal. Chem.*, 37, 167, 1965.

Table GC 6
ACRYLONITRILE MONOMER IN PLASTIC PACKAGING AND BEVERAGES BY HEADSPACE GAS CHROMATOGRAPHY

A gas-liquid chromatographic method for determining trace amounts of acrylonitrile in plastic containers and carbonated beverages using a nitrogen sensitive detector and headspace injection technique is described. The method is suitable for the determination of acrylonitrile at concentrations down to 0.1 mg Kg^{-1} in plastics and 0.005 mg Kg^{-1} in beverages.

Column packing	0.2% Carbowax 1500® on 60/80 mesh Carbopak C®
Temperature (°C)	70
Gas; flow rate (mℓ/min)	Nitrogen; 23
Column	
Length, m	3
Diameter, mm	2
Material	Glass
Detector	Alkali flame-ionization (rubidinum sulfate)

Gawell, G. B. M., *Analyst*, 104, 106, 1979.

Table GC 7
ACRYLONITRILE MONOMER IN ACRYLONITRILE CONTAINING POLYMERS AND FOOD SIMULATING SOLVENTS

For the solvents (heptane, 3% acetic acid, 8% and 50% ethanol) that have been in contact with polymer, a 5 $\mu\ell$ aliquot is analyzed by GC.

Column	Chromosorb 101®, 80/100
Temperature (°C)	130
Gas; flow rate (mℓ/min)	Helium; 25
Column	
Length, m	2
Diameter, mm I.D.	2
Material	SS
Detector	N-P flame photometric

For polymers (as 10% solution in dimethyl-acetamide), a 5 $\mu\ell$ aliquot is injected on a stainless steel precolumn (2 ft) containing Porapak QS® Column temperature is increased to 165°C.

Brown, M. E., Breder, Charles, V., and McNeal, T. P1, *Anal. Abstr.*, 36(6), 630, 1979.

Table GC 8
HEADSPACE DETERMINATION OF RESIDUAL ACRYLONITRILE IN ACRYLONITRILE-BUTADIENE-STYRENE RESINS AND MIGRATION INTO A SIMULATED FATTY FOOD

The resin was dissolved in DMF and water added to increase the concentration of acrylonitrile in the headspace vapor. After incubation at 80°C for 60 min, the vapor was analyzed by GC.

Column packing	10% Carbowax 20M® on Chromosorb WAW® 60/80
Temperature (°C)	95
Gas; flow rate (mℓ/min)	Nitrogen; 30
Column	
Length, m	3
Diameter, mm I.D.	2
Material	SS
Detector	FID

The migration of acrylonitrile was measured by storing sheets of the resin in olive oil for 10 days at 40°C. A portion of the oil was incubated in a sealed vial at 80°C for 1 hr and the headspace analyzed by GC.

Column packing	Chromosorb 102®, 100/120
Temperature (°C)	140
Gas; flow rate	Helium; 30
Column	
Length, m	2
Diameter, mm I.D.	2
Material	SS
Detector	Nitrogen specific

Di Pasquale, G., Di Iosio, G., Capaccioli, T., Gagliardi, P., and Verga, G. R., *Anal. Abstr.*, 36(6), 629, 1979.

Table GC 9
ALKOXYL SUBSTITUTION IN CELLULOSE ETHERS BY ZEISEL-GAS CHROMATOGRAPHY

Cellulose ethers are cleaved by reaction with hydriodic acid for 1 hr at 150°C using adipic acid as a catalyst. The alkyl iodide reactants are extracted with xylene and analyzed by gas chromatography.

Column packing	10% UCW-98
Temperature (°C)	100
Gas; flow rate (mℓ/min)	Helium; 20
Column	
Length, m	3
Diameter, mm I.D.	2
Material	SS
Detector	TC
Chart speed, cm/min	1

Compound	t(min)
CH_3I	3.3
C_2H_2I	4.3
C_3H_7I	5.3

Hodges, K. L., Kester, W. E., Wiederrich, D. L., and Grover, J. A., *Anal. Chem.*, 51, 2172, 1972.

Table GC 10
ALKYL-SUBSTITUTED ACRYLONITRILES

Column packing	P1
Temperature (°C)	70
Gas; flow rate (mℓ/min)	He; 65
Column	
Length, m	2
Diameter, mm	4
Material	n.a.
Detector	Katharometer

Compound	Retention time (min)
Acrylonitrile	3.30
Methacrylonitrile	4.75
cis-Tiglonitrile	8.53
α-Ethylacrylonitrile	10.03
trans-Angeliconitrile	21.75
cis-α-Methyl-β-ethylacrylonitrile	24.89
α-Propylacrylonitrile	30.07
trans-α-Methyl-β-ethylacrylonitrile	39.80
cis-α-Methyl-β-propylacrylonitrile	44.91
α-Butylacrylonitrile	49.03
trans-α-Methyl-β-propylacrylonitrile	59.84

Note: Column packing — P1 = 15% Silicone DC-550® on Celite 22® (80/100 mesh).

Kyazimova, G. P., Mekhtiv, S. I., and Mamedov, R. G., *Zh. Anal. Khim.*, 30, 1030, 1975.

Table GC 11
ALKYLCYANOACRYLATES

Column packing	P1
Temperature (°C)	170
Gas; flow rate (mℓ/min)	He; 40
Column	
Length, ft	6
Diameter, in. I.D.	0.178
Material	SS
Detector	FI

Compound	Retention volume (mℓ)
Methyl-2-cyanoacrylate	132.8
Ethyl-2-cyanoacrylate	144.8
n-Propyl-2-cyanoacrylate	211.2
sec-Butyl-2-cyanoacrylate	236.0
n-Butyl-2-cyanoacrylate	280.8
n-Amyl-2-cyanoacrylate	406.0
n-Hexyl-2-cyanoacrylate	585.2
n-Heptyl-2-cyanoacrylate	888.0
n-Octyl-2-cyanoacrylate	1200.0

Note: Column packing — P1 = 10% GE XE-60 (silicone gum nitrile) on Chromosorb P®.

Kulkarni, R. K., Johnson, E. C., and Wade, C. W. R., *Anal. Chem.*, 46, 749, 1974.

Table GC 12
ALLYL CHLOROFORMATE AND DIETHYLENE GLYCOL BIS (CHLOROFORMATE) IN DIETHYLENE GLYCOL BIS(ALLYL CARBONATE)

Column packing	P1	P1
Temperature (°C)	133	25
Gas; flow rate (mℓ/min)	N_2; 95	N_2; 95
Column		
Length, ft	6	6
Diameter, in. O.D.	3/16	3/16
Material	Glass	Glass
Detector	FI	FI

Compound	Retention time (min)	
Chloroform[a]	0.5	0.6
Allyl chloroformate		2.5
Diethylene glycol bis(chloroformate)	2.7	
Diethylene glycol bis(allyl carbonate)	16.5	

Note: Column packing — P1 = 5% Silicone gum rubber SE-30 on Anakrom AS® (80/90 mesh).

[a] Solvent

Gudzinowicz, B. J., *Anal. Chem.*, 37, 1051, 1965.

Table GC 13
AMINE ANTIOXIDANTS

Column packing	P1
Temperature (°C)	200, 220, or 250
Gas; Flow Rate (mℓ/min)	He; 50
Column	
Length, mm	2
Diameter, mm	3
Detector	FI

Compound	Standard	Relative retention time[a]	Response[b]	Column temperature, °C
DPPD	CPPD	1.25	1.35	250
CPPD	DPPD	0.80	0.74	250
IPDA	DA	2.83	0.82	200
PBNA	PANA	1.36	1.01	250
DMNA	TMPD	2.76	0.84	200
TMPD	DMNA	0.36	1.19	200
TDDM	DADM	2.08	1.87	220

[a] Retention time relative to standard.
[b] Area of amine per gram/area internal standard per gram.

DPPD — *N*,N′-diphenyl-*p*-phenylene diamine.
CPPD — *N*-cyclohexyl-N′-phenyl-*p*-phenylene diamine.
IPDA — *p*-Isopropoxydiphenylamine
DBNA — *N*-phenyl-2-naphthylamine
DMNA — *N*,N-dimethyl-2-napthylamine
TMPD — *N*,*N*,N′,N′-tetramethyl-*p*-phenylene diamine
TDDM — *N*,*N*,N′,N′-tetramethyl-*p*,p′-diaminodiphenyl methane
DA — Diphenyl amine
PANA — *N*-phenyl-1-naphthylamine
DADM — 4,4′-Diaminodiphenyl methane

Note: Column packing — P1 = 10% Apiezon L® on Chromosorb W® (30/60 mesh).

Tsurugi, J., Murakami, S., and Goda, K., *Rubber Chem. Technol.*, 44, 857, 1971.

Table GC 14
ANTIOXIDANTS

2,6-di-*tert*-butyl-4-methyl phenol (BHT) and 3-*tert*-butyl-4-hydroxyanisole (BHA) are amenable to analysis by gas chromatography. Lower detection limits are achieved by converting the polar compounds to less polar and more volatile derivatives as described below.

Dissolve 2 g BHA in 6 mℓ hexane and add 2.5 g trifluoroacetic anhydride. Seal in an ampoule and heat at 80°C for 1 hr. Cool to room temperature. Wash the contents with 0.1 *M* NaOH and then with water.

Column packing	P1	P2
Temperature (°C)	175	160
Gas; flow rate (mℓ/min)	10% methane in Argon;	40
Column		
Length, m	1.5	1.5
Diameter, mm O.D.	6	6
Material	Glass	Glass
Detector	FI	FI

Compound	**Relative retention time**	**Detection limit (ng)**
3-*tert*-butyl-4-hydroxy-anisole	1.00	
2-*tert*-butyl-4-methoxy-phenyl trifluoroacetate	0.68	0.5
2,6-Di-*tert*-butyl-4-methylphenol	1.21	

Note: Column packing — P1 = W AW DMCS (80/100 mesh).
P2 = 5% SE-30 on Chromosorb W AW DMCS® (80/100 mesh).

Dilli, S. and Robards, K., *J. Chromatogr.*, 133, 363, 1977.

Table GC 15
IMPURITIES IN BISPHENOL A

Dissolve 1 g of Bisphenol A® in a mixture of 3 mℓ pyridine and 3 mℓ of acetic anhydride. Heat on a steam bath for 20 min. Add 10 mℓ H_2O and 3 mℓ of concentrated HCl. Allow the mixture to cool and extract with 2 mℓ of benzene. Chromatograph a 3-μℓ aliquot of the solution containing acetylated Bisphenol A® and impurities.

Column packing	P1	P2	P3
Temperature (°C)	240	240	240
Gas; flow rate (mℓ/min)	He; 72	He; 94	He; 125
Column			
Length, in.	36	10	4
Diameter, in. O.D.	0.125	0.125	0.125
Material	SS	SS	SS
Detector	FI	FI	FI

Compound	Retention time (min)		
Benzene	0.5	0.5	—
Phenyl acetate	0.5	0.5	—
Codimer acetate[a]	16	2	—
o,p′-BPA acetate[b]	19	5	—
Bisphenol A acetate	38.5	41	—
BPX acetate[c]	142	—	14

Note: Column packing — P1 = 25% Lac-2R-446 + H_3PO_4 on Chromosorb W® (80/100 mesh).

[a] 4,4′-hydroxyphenyl-2,2,4-trimethylchroman.
[b] 2-(2-hydroxyphenyl)-2-(4-hydroxyphenyl)propane.
[c] 2,4-bis(α,α-dimethyl-4-hydroxybenzyl)phenol.

Gill, H. H., *Anal. Chem.*, 36, 2101, 1964.

Table GC 16
TRACE IMPURITIES IN 1,3-BUTADIENE

Column packing	P1
Temperature (°C)	25
Gas; flow rate (mℓ/min)	He; 35
Column[a]	
Length, ft	40
Diameter, in.	1/8
Material	SS
Detector	FI

[a] The column consists of two sections (P1A, 25 ft, and P1B, 15 ft) connected in series.

Compound	Retention time (min)
Methane	2.7
Ethane + ethylene	3.1
Propane	4.0
Acetylene	4.3
Propylene	4.4
Isobutane	5.6
Cyclopropane	6.1
Propadiene	6.7
n-Butane	7.0
Neopentane	7.8
Butene-1 + isobutylene	8.4
Methylacetylene	9.0
trans-Butene-2	9.6
cis-Butene-2	10.8
1,3-Butadiene	11.5
Isopentane	12.7
3-Methylbutene-1	13.5
n-Pentane	16.0
1,2-Butadiene	16.7
Ethylacetylene	18.5
2-Methylbutene-1	20.2
1,4-Pentadiene	21.5
trans-Pentene-2	22.3
Vinylacetylene	23.5
2-Methylbutene-2	25.8
2-Methylpentane	30.2
Isoprene	31.3
Dimethylacetylene	33.8

Note: Column packing — P1A = 20% Dibutyl maleate on Chromosorb P-NAW® (60/80 mesh). P1B = 10% Bis(2-methoxyethoxy)ethyl ether (BMEE) on Chromosorb P-NAW® (60/80 mesh).

Carson, J. W., Young, J. D., and Legé, G., *J. Chromatogr. Sci.*, 10, 737, 1972.

Table GC 17
CARBOXYLIC ESTERS

Difficultly saponifiable esters are hydrolyzed by an alkali fusion reaction. The alcohols from the saponification reaction are determined by gas chromatogrpahy. The method has been applied to the analysis of terephthalates, polyacrylates, polymethacrylates, halogenated esters (e.g., poly(methyl-α-chloroacrylate)), and copolymers. Weigh 1 to 4 mg of sample into a 11-mm micro-size platinum boat. Cover the sample with a 30 to 100 mole excess of the alkali fusion mixture consisting of potassium hydroxide containing 0.5 to 1% by weight sodium acetate. Heat the boat and contets for 30 min at 300° to 360°C. Trap the volatile alcohols and then sweep the products directly into the gas chromatograph by applying heat to the trap.

Column packing	P1	P1	P2
Temperature (°C)	75/3 min 24°/min to 200	75/3 min 24°/min to 200	75 to 100 1°/min
Column			
Length, ft	7	5	10
Diameter, in.	1/4	1/8	1/4
Material	SS	SS	Cu
Detector	TC	TC	TC

		Retention time (min)		
Polymer	**Compound**	**P1**	**P1**	**P2**
Poly(methyl methacrylate)	methanol	7		
Poly(methyl acrylate)	methanol	7		
Poly(methyl-α-chloroacrylate)	methanol		4.8	
Poly(ethyl-α-chloroacrylate)	ethanol		6.7	
Poly(isopropyl-α-chloroacrylate)	isopropanol		8.0	
Poly(vinyl pyrrolidone-co-ethyl acrylate)	ethanol			2.9

Note: Column packing — P1 = 10% DC-LSX-3-0295 trifluoropropyl silicone gum on acid-washed Chromosorb-P® (60/80 mesh).

Burson, K. R. and Kenner, C. T., *Anal. Chem.,* 41, 870, 1969.

Table GC 18
CHLOROSTYRENES AND ALKYLBENZENES

Column packing	P1	P2	P2
Temperature (°C)	150	150	100
Gas; flow rate (mℓ/min)	n.a.	n.a.	n.a.
Column			
Length, ft	12	12	12
Diameter, in. O.D.	3/16	3/16	3/16
Material	SS	SS	SS
Detector	TC	TC	TC

Compound	**Retention time (min)**		
Toluene	0.9	1.6	4.5
p-Chlorostyrene	6.6	10.5	
m-Chlorostyrene	7.4	11.3	
o-Chlorostyrene	8.1	12.1	
α-Chlorostyrene	9.0	13.0	
trans-β-Chlorostyrene	10.3	14.9	
cis-β-Chlorostyrene	11.3	16.4	
Ethylbenzene			8.0
p-Xylene			8.8
m-Xylene			9.7
o-Xylene			11.2

Note: Column packing — P1 = 10 pph Bentone-34® and 5 pph bis-(phenoxyphenyl) ether on Anakrom AS® (70/80 mesh).
P2 = 10 pph Bentone-34® and 5 pph dinonylphthalate on Chromosorb W® (60/100 mesh).

Raley, C. F. and Kaufman, J. W., *Anal. Chem.*, 42, 1371, 1968.

Table GC 19
ISOMERIC DIAMINOTOLUENES

Dissolve about 0.4 g of finely ground diaminotoluene (TDA) mixed-isomer sample in 2 mℓ of 10:1 (v/v) pyridine-methanol solution. Chromatograph 10-μℓ of the solution.

Column packing	P1
Temperature (°C)	215
Gas; flow rate (mℓ/min)	He; 100
Column[a]	
Length, ft	20
Diameter, in. O.D.	1/4
Material	Al

[a] The column consists of two sections (P1A, 14 ft, and P1B, 6 ft) connected in series.

Compound	B.P., °C	Relative retention time[a]
2,3-Diaminotoluene	255	0.707
3,4-Diaminotoluene	265 sub.	0.746
2,5-Diaminotoluene	273.4	0.820
2,6-Diaminotoluene	289	1.00
2,4-Diaminotoluene	292	1.12

Note: Column packing — P1A = 4% Carbowax 20M® on Chromosorb G-NAW® + 1% KOH (60/80 mesh).
P1B = 5% Siponate DS-10® on Chromosorb G-NAW® + 1% KOH (60/80 mesh).

[a] Relative to 2,6-diaminotoluene.

Willeboordse, F., Quick, Q., and Bishop, E. T., *Anal. Chem.*, 40, 1455, 1968.

Table GC 20
ISOMERIC DIAMINOTOLUENES, BIS-*N*-TRIFLUOROACETAMIDE DERIVATIVES

Procedure A

Weigh 60 to 100 mg of sample into a 5-mℓ serum bottle. Add 1 mℓ of trifluoroacetic acid and stopper the bottle with a rubber septum cap. Shake the mixture until the sample dissolves. Add 1 mℓ of trifluoroacetic anhydride with a syringe and let stand for 15 min. Chromatograph a 10-μℓ aliquot of the solution containing the bis-*N*-trifluoroacetamide derivatives on Column Packing P1.

Procedure B

Prepare the derivatives as described in Procedure A substituting chloroform for the trifluoroacetic acid. The derivatives are insoluble in chloroform and precipitate. Let stand for 15 min then open the bottle and evaporate the solvents and excess reagent using a stream of nitrogen. Dissolve the precipitate in 1 mℓ of acetone. Chromatograph a 5-μℓ aliquot of the solution on Column Packing P2.

Column packing	P1	P2
Temperature (°C)	200	225
Gas; flow rate (mℓ/min)	He; 80	He; 70
Column		
Length, ft	3	12
Diameter, in. O.D.	0.25	0.25
Material	Al	Al
Detector	TC	TC

Compound (bis-*N*-trifluoroacetamide derivative)	Relative retention time[a]	
2,3-Diamino toluene	0.387	0.524
3,4-Diamino toluene	0.465	0.623
2,6-Diaminotoluene	1.00[b]	1.00[c]
2,4-Diaminotoluene	1.25	1.13
2,5-Diaminotoluene	1.28	1.25

Note: Column packing — P1 = 2% Neopentylglycol succinate on Chromosorb G® (60/80 mesh).
P2 = 5% OS-138 six-ring *m*-polyphenyl ether on Chromosorb G®, AN-DMCS (100/120 mesh).

[a] Relative to 2,6 diaminotoluene.
[b] Retention time = 13.8 min.
[c] Retention time = 25 min.

Brydia, L. E. and Willeboordse, F., *Anal. Chem.*, 40, 110, 1968.

Table GC 21
TRACE IMPURITIES IN DIVINYLBENZENE

Column packing	P1	P2
Temperature (°C)	160	112
Gas; flow rate (mℓ/min)	He; 30	He; 30
Column		
Length, ft	20	20
Diameter	n.a.	n.a.
Material	n.a.	n.a.
Detector	FI	FI

Compound	Relative retention time[a]	
Benzene	0.08	0.04
Toluene	0.11	0.07
Ethylbenzene	0.16	0.14
Cumene	0.19	0.21
Ethyltoluene	0.23	0.28
Styrene	0.26	0.20
α-Methylstyrene	0.36	0.35
m-Diethylbenzene	0.37	0.41
p-Diethylbenzene	0.39	0.51
Vinyltoluene	0.42	0.41
m-Ethylvinylbenzene	0.62	0.71
p-Ethylvinylbenzene	0.65	0.76
Indene	0.74	0.57
m-Divinylbenzene	1.00	1.00
p-Divinylbenzene	1.08	1.09
1,2-Dihydronaphthalene	1.26	1.21
Naphthalene	2.08	1.22

Note: Column packing — P1 = 20% Carbowax 6000® on Gas Chrom RA®.
P2 = 20% Di-2-ethylhexylsebacate on Chromosorb W®.

[a] Relative to *m*-divinylbenzene.

Hannah, R. E., Cook, M. L., and Blanchette, J. A., *Anal. Chem.*, 39, 358, 1967.

Table GC 22
ETHANEDIOL IN FORMALDEHYDE-ETHANEDIOL POLYMERS

Weigh 200 to 300 mg of pulverized polymer into a flask and add 1.5 mℓ acetic anhydride and one drop of perchloric acid. Reflux for 15 min. Cool to room temperature and let stand for 15 min. Chromatograph 1-μℓ of the acetylated diol solution.

Column packing	P1
Temperature (°C)	130—200 4°/min
Gas; flow rate (mℓ/min)	He; 30
Column	
Length, m	4
Diameter, mm I.D.	3
Material	SS
Detector	FI

Compound	Retention time (min)
Acetic acid	9
Acetic anhydride	14.5
Ethanediol diacetate	19
1,3-Propanediol diacetate	23
1,4-Butanediol diacetate	27.5
1,5-Pentanediol diacetate[a]	31

Note: Column packing — P1 = 10% Polyglycol 20,000® on Chromosorb W-AW-DMCS®.

[a] Internal standard.

Fischer, R., *Z. Anal. Chem.*, 251, 172, 1970.

Table GC 23
ETHYL ACRYLATE IN ACRYLIC EMULSIONS

Add 50-mℓ of emulsion to a 500-mℓ boiling flask and dilute with 125 mℓ of water. Add 3 mℓ of toluene, 20 mg of hydroquinone and a small amount of antifoam. Reflux the mixture until 3 mℓ of organic layer are collected in the receiver. Chromatograph a 0.5 μℓ sample of the distillate.

Column packing	P1
Temperature (°C)	87
Gas; flow rate (mℓ/min)	He; n.a.
Column	
Length, ft	6
Diameter	n.a.
Detector	TC

Compound	Retention time (min)
Ethyl acrylate	4.8
Toluene	7.2

Note: Column packing — P1 = 30% silicone stopcock grease on Celite 545 AW® (30/60 mesh).

Tweet, O. and Miller, W. K., *Anal. Chem.*, 35, 852, 1963.

Table GC 24
ETHYL ACRYLATE-METHYL METHACRYLATE IN ACRYLIC EMULSION

Reflux 100 g of acrylic emulsion and 5 g of benzene for 30 min in 500-mℓ boiler connected to an azeotropic distillation receiver. Inject 10 μℓ of the organic phase in the distillation receiver onto the chromatographic column. The reproducibility between 100 and 1000 ppm is about 5% for ethyl acrylate and 10% methyl methacrylate.

Column packing	
Precolumn	P1
Analytical column	P2
Temperature (°C)	60
Gas; flow rate (mℓ/min)	H_2; 40
Precolumn	
Length, m	0.3
Diameter, mm I.D.	4
Material	Cu
Analytical column	
Length, m	4
Diameter, mm I.D.	4
Material	Cu
Detector	Thermistor

Compound	Relative retention time
Acetone	1.00
Ethanol	2.07
Ethyl acrylate	2.89
Methyl methacrylate	3.33

Note: Column packing — P1 = 30% PEG 200 on Chromosorb P® (60/80 mesh).
P2 = PPG 400 plus 5% Bentone 34® on Chromosorb P® (60/80 mesh).

Danoy, J. and Deneubourg, C., *J. Chromatogr.*, 56, 126, 1971.

Table GC 25
ETHYLENIMINE OLIGOMERS

Column packing	P1
Temperature (°C)	150—240 12°/min
Gas; flow rate (mℓ/min)	N_2; 25
Column	
Length, m	1.0
Diameter, cm	0.4
Detector	n.a.

Compound	Approximate elution temperature, °C
Ethylenimine	153
Dimer	155
β-Picoline[a]	157
Trimer	166
Quinoline[a]	180
Branched tetramer	196
Linear tetramer	204
Isomeric pentamer[b]	214
Isomeric pentamer[c]	221
Isomeric pentamer[d]	230

Note: Column packing — P1 = 10% Apiezon L® on Celite-545® treated with 10% alkali.

[a] Internal standard.

[b] Probable structure:

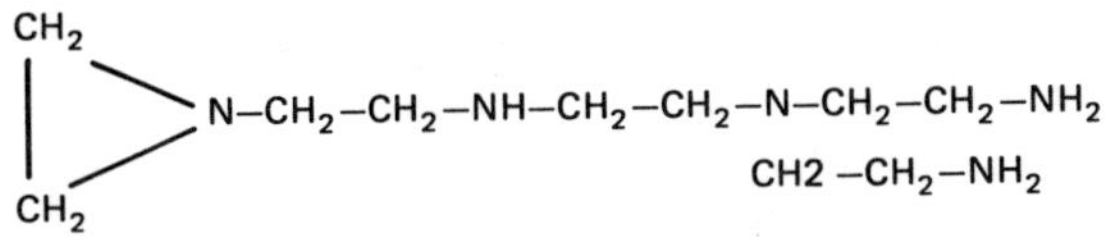

[c] Probable structure:

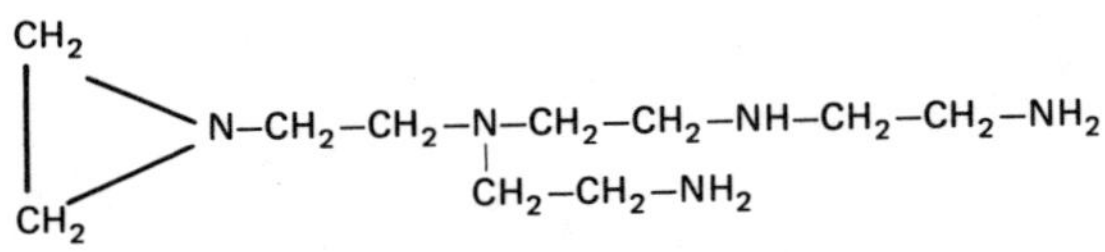

[d] Probable structure:

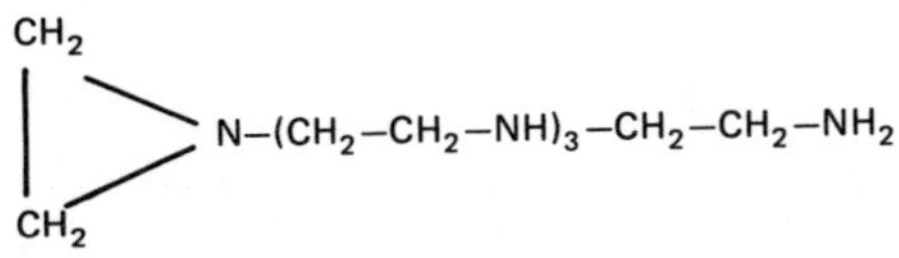

Gromova, G. L., Berezkin, V. G., Gembitskii, P. A., and Zhuk, D. S., *Polym. Sci. USSR,* 18(1), 279, 1976.

Table GC 26
CHEMICAL CHANGES OF FURFURYLIDENE ACETONES DURING THERMAL PROCESSING

The test tube method, followed by gas chromatography, has been used to analyze the volatile products of the thermochemical conversion of furfurylidene- and difurfurylidene acetones in the range of 200 to 500°C as well as those of a series of compounds clarifying some of the processes which take place during structuration. All of these compounds have been found to liberate water, oxides of carbon, hydrogen and lower hydrocarbons

Column	Porapak Q®
Column temperature	90°C
Carrier	Argon, 20 mℓ/min
Detector	Thermal conductivity

Korshak, V. V., Tseithin, G. M., Khomutov, V. A., Gonsales, Kh. P., and Atrushkevich, A. A., *Polym. Sci. USSR,* 21(61), 1979.

Table GC 27
RESIDUAL MONOMERS IN LATEX

Dissolve 1 g of latex in a suitable solvent such as acetic anhydride, 2,2-dimethoxypropane, or cyclohexanone. Acetic acid is formed when acetic anhydride is the solvent and methanol and acetone are produced when 2,2-dimethoxypropane is used. The monomer detection limits vary from about 100 ppm to 500 ppm.

Column packing	P1	P1
Temperature (°C)	150	200
Gas; flow rate (mℓ/min)	He; 85	He; 100
Column		
Length, ft	10	10
Diameter, in. O.D.	1/4	1/4
Material	Cu	Cu
Detector	TC	TC

Compound	N.R.V. (mℓ[a])	ART (min)[b]	N.R.V. (mℓ[a])	ART (min)[b]
Acrylonitrile	308	4		
n-Butyl acrylate	663	8.4		
Ethyl acrylate	280			
2-Ethylhexyl acrylate			865	9
Isobutyl acrylate	500			
Methyl acrylate	238			
Methyl methacrylate	272			
Styrene			430	4.8
Vinyl acetate	193			
Vinyl chloride	45	1.2		
Vinyl propionate	272			
Vinylidene chloride	90	2.2		
Acetic acid				
Acetic acid		10.8		
Acetone		23.8		0.8
Cyclohexanone				
Methanol		21.7		0.8
Water		4.4		

Note: Column packing — P1 = 30% polyethylene glycol E-6000 on Chromosorb P® (30/60 mesh).

[a] Net retention volume corrected to 30°C but uncorrected for pressure drop.
[b] Approximate retention time.

Wilkinson, L. B., Norman, C. W., and Beuttner, J. P., *Anal. Chem.*, 3, 1759, 1964.

Table GC 28
OLEFINS AND AROMATIC HYDROCARBONS

Column packing		P1	P2	P1	P2
Temperature (°C)		58	129	58	129
Gas; flow rate (mℓ/min)		N_2; 20.1	N_2; 20.1	N_2; 20.1	N_2; 20.1
Column					
Length, m		6	6	1.4	1.4
Diameter, mm I.D.		4	4	3.6	3.6
Material		Al	Al	Glass	Glass
Detector		FI	FI	FI	FI
Compound	**B.P.°C**	**Relative retention time[a]**			
1,3-Butadiene	−2.6	0.35	0.65		
n-Butane	−0.5	0.29	0.31		
2-Methylbutadiene-1,3	35.0	1.10	1.76		
n-Pentane	36.0	1.00	1.00		
cis-2-Pentene	37.0	1.14	1.24		
2-Methyl-2-butene	38.6	1.20	1.38		
Cyclopentadiene	41—42	1.36	—		
4-Methyl-trans-2-pentene	53.9	2.18	2.35		
2,3-Dimethyl-1,3-butadiene	69—70	3.29	6.00		
n-Hexane	68.9	3.15	3.00		
Benzene	80.08			0.06	0.18
Cyclohexane	81.0			0.06	0.06
Toluene	110.6			0.24	0.41
Ethylbenzene	135.5			0.43	0.70
o-Xylene	144.1			0.47	0.94
m-Xylene	139.3			0.46	0.83
p-Xylene	138.0			0.46	0.76
Styrene	145.0			0.47	0.75
Phenylacetylene	142—144			0.56	0.60
Isopropylbenzene	152.0			0.57	1.01
Mesitylene	164.8			0.86	1.57
tert-Butylbenzene	168.0			0.91	1.59
1-Methyl-4-isopropylbenzene	177.0			1.11	1.71
n-Decane	174.0			1.00	1.00
Naphthalene	218.0			3.74	5.98
Phenyl-*tert*-butylacetylene	72°/2 mm			2.56	8.59
Benzo-*tert*-butylcyclobutene	57°/1 mm			1.90	3.77
p-Di-tert-butylbenzene	236.5			6.69	17.23
n-Tridecane	234.0			5.90	7.66
1-Methyl-naphthalene	241.0			5.65	10.30
2-Methyl-naphthalene	240—242			6.74	10.76
Hexamethylbenzene	264.0			9.04	18.69
Diphenyl	254—255			14.93	17.38
n-Pentadecane	270.5			20.81	32.20

Note: Column packing — P1 = 25% Sodium tetraphenylborate on Chromosorb W® (30/60 mesh).
P2 = 25% thallium (I) tetraphenylborate on Chromosorb W® (30/60 mesh). The $T1B(C_6H_4)_4$ is prepared by precipitation from a concentrated solution of thallium (I) sulfate with a dilute solution of sodium tetraphenylborate, followed by filtration and washing.

[a] Relative to *n*-pentane on P1; *n*-decane on P2.

Baiulescu, G. E. and Llie, V. A., *Anal. Chem.*, 46, 1847, 1974.

Table GC 29
PHENOLIC ANTIOXIDANTS

Column packing	P1	P2	P3	P2
Temperature (°C)	137	80—140 5°/min	100—250 10°/min	50—250 10°/min
Gas; flow rate (mℓ/min)	N_2; 30	N_2; 30	N_2; 30	N_2; 30
Column				
Length, cm	200	100	100	100
Diameter, cm	0.4	0.4	0.4	0.4
Material	SS	SS	SS	SS
Detector	n.a.	n.a.	n.a.	n.a.

	Approximate retention time (min)			
Compound	**P1**	**P2**	**P3**	**P2**
o-Cresol	12			
p-Cresol	15.8			
m-Cresol	17.5			
Phenol		2.5	2.8	
tert-Butoxybenzene		2.5		
m-*tert*-Butylphenol		6.6		
p-*tert*-butylphenol		6.9		
2,6-di-*tert*-Butylphenol		9.2		
2,4-di-*tert*-Butylphenol		10.8		
2,4,6-tri-*tert*-Butylphenol		12.5		
2-*tert*-Butyl-4-methyl-6-styrene-sub. phenol				7
2-Styrene-sub. phenol			8.6	
4-Styrene-sub. phenol			10	
2,6-di-Styrene-sub. phenol			12.2	
2,4-di-Styrene-sub. phenol			13.8	
2,4,6-tri-Styrene-sub. phenol			15.8	
2,6-di-*tert*-Butyl-*p*-cresol				21(158°)
2,4,6-tri-*tert*-Butylphenol				22(171°)
Phenyl-β-naphthylamine				29.5(246°)
2,2′-Methylene-bis(4-methyl-6-*tert*-butylphenol)				28(232°)

Note: Column packing — P1 = 2% Neopentyl glycol succinate and 2% Bentone-245® on Chromosorb G-AW-DMCS® (100/120 mesh).
P2 = 0.1% Neopentyl glycol succinate and 4% methylvinyl-silicone CKTB on Porochrom-1-AW-HMDS® (0.25—0.38 mm).
P3 = 2% Neopentyl glycol succinate on Porochrom-1-AW-HMDS® (0.38—0.5 mm).

Styskin, E. L. and Gurvich, Y. A., *J. Chromatogr.*, 77, 11, 1973.

Table GC 30
FREE PHENOL AND FREE FORMALDEHYDE IN PHENOLIC RESINS

Weigh about 10 g of resin together with a known amount of 1-butanol (internal standard for formaldehyde) and *m*-cresol (internal standard for phenol) and dilute with an equal volume of water. Divide the solution into two fractions. Acidify one fraction with concentrated HCl or H_2SO_4. Filter off the resin solids and inject 2 to 6 μℓ of the aqueous solution for HCHO determination (5 to 0.3%). Add 10-mℓ of ether to the second fraction and neutralize, while stirring, with acid. Inject 6 to 25 μℓ of the ether extract for the phenol determination (1.8 to 0.6%).

Column packing	P1	P2
Temperature (°C)	130	130
Gas; flow rate (mℓ/min)	He; 120	He; 59
Column		
Length, ft	10	16
Diameter, in. O.D.	1/4	1/4
Material	Cu	Cu
Detector	TC	TC

Compound	Retention time (min)	
Formaldehyde	—	1.2
Methanol	0.4	1.9
Water	0.3	2.1
Ether	0.5	—
1-Butanol	0.8	4.4
Phenol	3.0	—
m-Cresol	5.0	—

Note: Column packing — P1 = 10% G.E. silicone SF-96 on Fluoropak®.
P2 = 10% sucrose octaacetate on Teflon 6®.

Stevens, M. P. and Percival, D. F., *Anal. Chem.*, 36, 1023, 1964.

Table GC 31
SUBSTITUTED PHENOLS

Column packing	P1
Temperature (°C)	230
Gas; flow rate (mℓ/min)	He; 22
Column	
Length, m	1
Diameter, mm	1.5
Material	Glass
Detector	TC

Compound	Relative retention time[a]
Phenol	0.33
3- and 4-Methyl phenol	0.44
2,4- and 2,5-Dimethyl phenol	0.52
2,3- and 3,5-Dimethyl phenol	0.62
2,3,6-Trimethyl phenol	0.69
2,3,5-Trimethyl phenol	0.82
3,4,5-Trimethyl phenol	1.08
2,3,5,6-Tetramethyl phenol	1.08
Pentamethyl phenol	2.11
Pyrocatechol	1.00
3-Methyl pyrocatechol	1.15
4-Methyl pyrocatechol	1.35
Hydroquinone	1.66
Resorcinol	1.83
2-Methyl resorcinol	1.68
4-Methyl resorcinol	2.02
4-Ethyl resorcinol	2.55
3-*tert*-Butyl pyrocatechol	2.03
4-*tert*-Butyl pyrocatechol	2.59
3,5-Di-*tert*-butyl pyrocatechol	3.84

Note: Column packing — P1 = 3% Carbowax 20M® on Porapak P® (80/100 mesh).

[a] Relative to pyrocatechol.

Kusy, V., *J. Chromatogr.,* 57, 132, 1971.

Table GC 32
IMPURITIES IN PHTHALIC ANHYDRIDE

Weigh 1.0 g of phthalic anhydride into a 250-mℓ beaker. Add 50 mℓ of benzene and warm to dissolve the sample. Cool to room temperature and add 15 mℓ of an alcohol-free ethereal solution of diazomethane containing about 3 g diazomethane in 175 mℓ of solution. Evaporate the solution to about 30 mℓ in a well ventilated hood. Transfer to a 50-mℓ volumetric flask. Dilute to volume with benzene and chromatograph a 2 μℓ of the solution containing the methyl esters and other impurities.

Column packing	P1
Temperature (°C)	185
Gas; flow rate (psi) inlet pressure)	Ar; 7.5
Column	
Length, ft	8
Diameter, mm	6
Material	n.a.
Detector	β-Ionization

Compound	Retention time (min)
Methyl benzoate	5.4
Methyl toluate	6.2
o-Xylene	7.4
Naphthalene	8.3
Phthalic anhydride	10.8
Phthalide	13.3
Napthoquinone	14.8

Note: Column packing — P1 = 25% silicone grease on Chromosorb W® (80/100 mesh).

Moore, C. E. and Meinstein, S., *Anal. Chem.*, 34, 1503, 1962.

Table GC 33
TRACE IMPURITIES IN PHTHALIC ANHYDRIDE

Dissolve 2.5 g of phthalic anhydride in o-dichlorobenzene in a 50-mℓ volumetric flask. If the solution is hazy, add 5 mℓ of acetone. Dilute to the mark with o-dichlorobenzene.

Column packing	P1
Temperature (°C)	220
Gas; flow rate (mℓ/min.)	He; 60
Column	
Length, ft	6
Diameter, in.	1/4
Material	SS
Detector	FI

Compound	Retention time (min)	Relative retention[a]
Acetone	1.1	0.24
Maleic anhydride	2.4	0.52
o-Dichlorobenzene	4.6	1.00
Benzoic acid	6.2	1.35
Naphthalene	8.1	1.83
Phthalic anhydride	11.4	2.48
1,4-Naphthoquinone	16.3	3.54

Note: Column packing — P1 = 30% Silicone SF96 on Chromosorb W® (60/80 mesh).

[a] Relative to o-dichlorobenzene.

Trachman, H. and Zucker, F., *Anal. Chem.*, 36, 269, 1964.

Table GC 34
ESTER PLASTICIZERS

Dissolve about 1 g of polymer in 100 mℓ of tetrahydrofuran. Allow the insoluble material to settle out and inject an aliquot of the clear solution.

Column packing	P1	P2
Temperature (°C)	100/4 min then 8°/min to 330	235
Gas; flow rate (psi)	He; 60	He; 10 (110 mℓ/min)
Column		
Length, ft	6	0.75
Diameter, in. O.D.	1/8	1/4
Material	SS	SS
Detector	FI	FI
Literature	1	2

Compound	Relative retention time	Retention time (min)
Triethyl citrate	62.5	
Methylphthalylethyl glycolate	73.6	
Ethylphthalyethyl glycolate	76.4	
Dibutyl phthalate	76.9	1.2
Dibutyl sebacate	85.9	
Acetyltributyl citrate	89.1	
Butylphthalylbutyl glycolate	91.3	
Butylbenzyl phthalate	92.8	
Trioctyl phosphate	92.8	
Di-2-ethylhexyl adipate	95.0	
Di-2-ethylhexyl phthalate (DOP)	100.0	
Di-2-ethylhexyl azelate	105.4	
Di-2-ethylhexyl sebacate	108.7	
Di-*n*-decyl phthalate	120.0	
Benzylbutyl phthalate		3.4
Dibenzyl phthalate		9.6

Note: Column packing — P1 = 10% UCW-98 on Diatoport A® (60/80 mesh).
P2 = 25% silicone grease on Embacel® (80/125 mesh).

[a] Relative to di-2-ethylhexylphthalate (DOP).

Krishen, A., *Anal. Chem.,* 43, 1130, 1971.

Cook, C. D., Elgood, E. J., Shaw, G. C., and Solomon, D. H., *Anal. Chem.,* 34, 1177, 1962.

Table GC 35
ESTER PLASTICIZERS

Extract the sample with methylene dichloride. Concentrate the extract taking care to avoid complete removal of the solvent. Dilute to 25 mℓ with methyl isobutyl ketone.

Column packing	P1
Temperature (°C)	185
Gas flow rate (mℓ/min)	N_2; 25
Column	
Length, m	2
Diameter, in.	1/8
Material	SS
Detector	FI

Compound	Relative retention time[a]
Triacetin	0.46
Diacetin	0.74
Dimethyl sebacate	0.89
Dimethyl phthalate	1.00
Diethyl phthalate	1.28

Note: Column packing — P1 = 5% Antarox CO-990® on Chromosorb G-AW-DMCS® (80/100 mesh).

[a] Relative to dimethylphthalate.

Tunstall, F. I. H., *Anal. Chem.,* 42, 542, 1970.

Table GC 36
ESTER PLASTICIZERS

Weigh 2 to 3 g of sample (nitrocellulose, vinyl, or acrylic-type lacquer) into a 25-mℓ flask. Add 1 mℓ of acetone and, with rapid stirring, 3 mℓ of CCl_4 and 3 mℓ of petroleum ether. Filter and chromatograph a 10-20 μℓ of the filtrate.

Column packing	P1
Temperature (°C)	210—290
	4°/min
Gas; flow rate (mℓ/min)	He; 120
Column	
Length, ft	6
Diameter, in.	1/4
Material	Cu
Detector	TC

Compound	Relative retention time[a]
Dimethyl phthalate	0.16
Diethyl phthalate	0.26
Dibutyl phthalate	0.67
Dibutyl sebacate	1.00
Butyl benzyl phthalate	1.32
Di-(2-ethylhexyl)phthalate	1.65
Tricresyl phosphate	1.78—1.97
Dioctyl sebacate	2.20

Note: Column packing — P1 = 20% silicone grease on Chromosorb W®.

[a] Relative to dibutyl sebacate.

Esposito, G. C., *Anal. Chem.,* 35, 1439, 1963.

Table GC 37
PLASTICIZERS AND PLASTICIZER HYDROLYSIS PRODUCTS

Column packing	P1
Temperature (°C)	100 for 4 min 8°/min to 330
Gas; flow rate (mℓ/min)	He; 15
Column	
Length, m	1.8
Diameter, mm	3.2
Detector	FI

Compound	Relative retention time[a]
Phthalates	
Dibutyl	76.9
Butyl benzyl	92.8
Di(2-ethylhexyl)	100.0
Di-*n*-decyl	120.0
Adipates	
Di(2-ethylhexyl)	95.0
Azelates	
Di(2-ethylhexyl)	105.4
Citrates	
Triethyl	62.5
Acetyltributyl	89.1
Sebacates	
Dibutyl	85.9
Di(2-ethylhexyl)	108.7
Glycolates	
Methylphthalylethyl	73.6
Ethylphthalylethyl	76.4
Butylphthalylbutyl	91.3
Phosphates	
Trioctyl	92.8
Components	
Methanol	0.034
Butanol	1.82
Pentanol	3.98
Hexanol	8.04
Hepthanol	15.5
2-Ethylhexanol	21.0
Octanol	24.5
Dimethyl adipate	37.0
Decanol	40.0
Dimethyl-*o*-phthalate	51.1
Dodecanol	53.0
Tetradecanol	64.3
Methyl palmitate	76.1
Methyl stearate	84.8

Note: Column packing — P1 = 10% Silicone Gum Rubber UCW 98® on Diaport S® (60/80 mesh).

[a] Relative to di(2-ethylhexyl)phthalate.

Krishen, A., *Anal. Chem.*, 43, 1131, 1971.

Table GC 38
RETAINED SOLVENTS IN PLASTIC FILMS AND LAMINATES

Crumple about 250 cm^2 of material loosely and place in a 1-lb Kilner jar. Seal the jar with a metal lid which has a brass 1/4-inch pipe union with a rubber septum sealed with Araldite® into a hole drilled through the lid. Inject 3 $\mu\ell$ of butyl acetate (internal standard) into the jar through the septum. Place the jar in an oven at 100°C for 90 min. Remove the jar and immediately chromatograph 2 mℓ of the atmosphere within.

Column	P1
Temperature (°C)	80
Gas; flow rate (mℓ/min)	n.a.
Column	
Length, ft	5
Diameter, in.	1/4
Material	Glass
Detector	FI

Compound	**Retention time (min)**
Ethanol	0.6
Methyl ethyl ketone	1.2
Butanol	2.6
Toluene	3.2
Butyl acetate	4

Note: Column packing — P1 = 9% Silicone oil and 3% UCON HB2000 on Chromosorb W®.

Davies, J. T. and Bishop, J. R., *Analyst,* 96, 55, 1971.

Table GC 39
POLYAMIDES BY HYDROLYSIS AND DERIVATIZATION

Weigh 0.1 g of polyamide into a glass tube and add 10 mℓ of 6N HCl. Seal the tube and place in an oven at 130°C. Hydrolysis of nylon 6 requires 2 hr, nylon 66 requires 4 hr, and nylon 11 and 13 require 8 hr. Transfer the hydrolysate to a 30-mℓ evaporating dish and evaporate to dryness. Add 40 mℓ of anhydrous 5% HCl in methanol and 1 mℓ of 2,2-dimethoxypropane and reflux for 3 hr. Remove the solvents under reduced pressure. Add 1.0 mℓ of trifluoroacetic anhydride and 1.0 mℓ of methylene chloride to the residue and store at room temperature for 2 hr. Chromatograph a 10-μℓ aliquot of the solution.

Column packing	P1	P1	P1
Temperature (°C)	140—220 4°/min	200	220
Gas; flow rate (mℓ/min)	He; 80	He; 80	He; 80
Column			
Length, m	2	2	2
Diameter, mm	3	3	3
Material	SS	SS	SS
Detector	TC	TC	TC

	Retention time (min)		
Compound	**P1**	**P1**	**P1**
Di-*N*-trifluoroacetylated diamine			
Trimethylenediamine	18.0		4.2
Tetramethylenediamine	19.9		5.3
Pentamethylenediamine	21.5		6.6
Hexamethylenediamine	23.0		8.1
Heptamethylenediamine	24.9		9.9
Octamethylenediamine	27.9		12.9
Nonamethylenediamine	30.8		15.5
Decamethylenediamine	35.8		20.6
Dodecamethylenediamine	49.2		33.7
N-Trifluoroacetylated amino acid methyl ester			
Glycine	3.1		0.7
β-Alanine	4.0		0.9
γ-Amino-*n*-butyric acid	6.8		1.6
δ-Amino-*n*-valeric acid	9.7		2.4
ε-Amino-*n*-caproic acid	12.7		3.3
η-Amino-*n*-caprylic acid	16.7		6.2
κ-Amino-*n*-undecanoic acid	22.5		14.8
λ-Amino-*n*-lauric acid	24.7		20.8
Dibasic acid dimethyl ester			
Succinic acid	1.7		
Glutaric acid	2.6		
Adipic acid	3.8		
Pimelic acid	5.2		
Suberic acid	6.8		
Azelaic acid	8.7		
Sebacic acid	10.6		
Dodecanedioic acid	14.8		

Note: Column packing — P1 = 5% Neopentyl glycol succinate polyester on Celite 545® (80/100 mesh).

Mori, S. and Furusawa, M., *Anal. Chem.*, 42, 138, 1970.

Table GC 40
LINEAR POLYAMIDES AND COPOLYAMIDES BY ALKALI FUSION AND ESTERIFICATION

Linear polyamides and copolyamides are decomposed by alkali fusion using 0.1 g polymer and 1 g of potassium hydroxide containing 0.5 to 1% sodium acetate. Place the sample and fusion mixture in a 9 mm O.D. borosilicate tube and seal under reduced pressure. Heat to 300°C, cool and add water to the reaction product. Extract the diamines with two 30-mℓ portions of *n*-butanol. Reduce the volume to 5 mℓ and chromatograph 1 μℓ of the solution. Acidify the aqueous solution with HCl and evaporate to dryness. Add 10 mℓ of boron trifluoride-methanol reagent and reflux for 1 hr. Extract the diesters twice with toluene. Dry over anhydrous $MgSO_4$, evaporate to approximately 5 mℓ, and chromatograph 1 μℓ of the solution.

Column packing	P1	P2
Temperature (°C)	200	200
Gas; flow rate	n.a.	n.a.
Column		
Length, m	2	2
Diameter, mm I.D.	6.35	6.35
Material	SS	SS
Detector	FI	FI

Compound	Retention time
Dimethyl ester	
Adipic acid	4.7
Suberic acid	9
Azelaic acid	12.6
Sebacic acid	17.5
1,6-Diaminohexane	13

Note: Column packing — P1 = 1% Apiezon L® on glass beads (60/80 mesh).
P2 = 5% Diethylene glycol adipate on Chromosorb W-AW DMCS®.

Glading, G. J. and Haken, J. K., *J. Chromatogr.*, 157, 404, 1978.

Table GC 41
POLYAMIDES BY HYDROLYSIS AND TRIMETHYLSILYATION

Weigh 0.1 g of polyamide into a glass tube and add 10 mℓ of 6N HCl. Seal the tube and place in an oven at 130°C. Hydrolysis of nylon 6 requires 2 hr, nylon 66 requires 4 hr, and nylon 11 and 13 require 8 hr. Transfer the hydrolysate to a 30-mℓ evaporating dish and evaporate to dryness. Dissolve 10 mg of the dry residue in 0.15 mℓ acetonitrile in a 10-mℓ flask and add 0.03 Mℓ of triethylamine and 0.15 mℓ of bis(trimethylsilyl) acetamide. Heat the contents for 30 min at 90°C. Add an additional 0.15 mℓ of acetonitrile and of bis(trimethylsilyl)acetamide and repeat the heating cycle. Dilute to 2-mℓ with acetonitrile and chromatograph 10-μℓ of the final solution.

Column packing	P1	P1	P1
Temperature (°C)	120—220 4°C/min	160	220
Gas; flow rate (mℓ/min)	He; 40	He; 40	He; 40
Column			
Length, m	2	2	2
Diameter, mm	3	3	3
Material	SS	SS	SS
Detector	TC	TC	TC

Component (TMS derivative)	**Retention time (min)**		
Diamine			
Tetramethylene diamine	8.0	3.1	
Pentamethylene diamine	10.0	4.5	1.0
Hexamethylene diamine	12.5	7.1	1.2
Heptamethylene diamine	15.0	11.0	1.5
Octamethylene diamine	17.2	16.1	1.8
Nonamethylene diamine	19.5	23.5	2.4
Decamethylene diamine	21.4		3.0
Dodecamethylene diamine	24.8		5.0
Amino acid			
Glycine	2.2		
β-Alanine	4.5		
γ-Amino-*n*-butyric acid	6.4		0.7
δ-Amino-*n*-valeric acid	8.5		0.8
ε-Amino-*n*-caproic acid	10.4	5.0	1.0
η-Amino-*n*-caprylic acid	15.0	11.7	1.5
κ-Amino-*n*-undecanoic acid	21.7		3.5
λ-Amino-*n*-lauric acid	23.2		4.2
Dibasic acid			
Succinic acid	5.3		1.8
Glutaric acid	7.1		1.6
Adipic acid	9.4		3.9
Pimeric acid	11.9		5.8
Suberic acid	14.0	8.9	1.4
Azelaic acid	16.7	13.5	1.8
Sebacic acid	18.2	20.8	2.3
Dodecanedioic acid	22.2		3.9

Note: Column packing — P1 = 5% Neopentyl glycol succinate polyester on Celite 545® (80/100 mesh).

Mori, S. and Furusawa, M., *Anal. Chem.,* 42, 959, 1970.

Table GC 42
CYCLIC MONOMER AND OLIGOMERS IN POLYAMIDES

Extract 10 g of polyamide with 300 g of ethanol under reflux for 20 hr. Evaporate the solution to dryness. Add 100 to 200 mg of the extraction residue to a suspension of 1.0 g $LiAlH_4$ in 30 mℓ of tetrahydrofuran. Reflux for 25 hr in a nitrogen atmosphere. Destroy the excess $LiAlH_4$ by the dropwise addition of a water-THF solution. Add 150 mℓ of 15% NaOH and extract the cycloamine derivatives into diethyl ether. Concentrate the ether extract to 10 mℓ and chromatograph 2-μℓ of the solution.

Column packing P1	P1	P1	P1	P1	P2
Temperature (°C)	35	100	100—200 10°/min	130—210 10°/min	130—270 15°/min
Gas; flow rate (mℓ/min)	N_2; 40	N_2; 40	N_2; 40	N_2; 40	N_2; 40
Column					
Length, m	2	2	2	2	2
Diameter, mm I.D.	3	3	3	3	3
Material	SS	SS	SS	SS	SS
Detector	FI	FI	FI	FI	FI
Compound cyclic amine derivative			**Retention time, min**		
Nylon 6 monomer	1.5				
dimer		5.8		1.5	3.0
trimer				7.9	7.5
tetramer				21.5	11.5
pentamer					38.6
hexamer					66.0
Nylon 12 monomer		2.2	2.2		1.4
dimer			22.6		11.8
trimer					38.6
Caprolactam			4.0		
6-aminohexanol			5.5		
Lauryl lactam		9.2			

Note: Column packing — P1 = 5% Neopentyl glycol succinate polyester + 3% KOH on Celite 545® (80/100 mesh).
P2 = 5% SE-30 + 3% KOH on Celite 545® (80/100 mesh).

Mori, S. and Furusawa, M., *Anal. Chem.*, 42, 661, 1970.

Table GC 43
DICARBOXYLIC ACIDS IN POLYAMIDES, METHYL ESTERS

Reflux 0.05 to 0.2 g sample with 25 mℓ 6N HCl for 24 hr. Transfer the hydrolysate to a 100-mℓ beaker and evaporate to dryness. Add 2 mℓ of azelaic acid (internal standard) and 10 mℓ BF_3-methanol solution and reflux for 2 hr. Transfer to a separatory funnel containing 50 mℓ of water. Extract with two 30-mℓ portions of diethyl ether. Evaporate the ether extract to about 0.5 to 1.0 mℓ total volume.

Column packing	P1
Temperature (°C)	100—220 5°C/min
Gas; flow rate	n.a.
Column	
Length, ft	6
Diameter, in.	1/4
Material	SS
Detector	FI

Compound	Relative retention time[a]
Dimethyl adipate	0.3
Dimethyl azelate	1.0
Dimethyl sebacate	1.2
Dimethyl dodecanedioate	1.7

Note: Column packing — P1 = 5% diethylene glycol adipate on Chromosorb WAW® (60/80 mesh).

[a] Relative to internal standard dimethylazelate.

Anton, A., *Anal. Chem.,* 40, 1116, 1968.

Table GC 44
RESIDUAL CAPROLACTAM IN POLYCAPROLACTAM (NYLON 6)

Dissolve about 0.2 g of polymer in 10 mℓ of formic acid. Inject a 1-μℓ aliquot for 0.1 to 10% residual caprolactam in the polymer.

Column packing	P1
Temperature (°C)	200
Gas; flow rate (mℓ/min)	N_2; 24
Column	
Length, m	0.8
Diameter, mm I.D.	4.0
Material	SS
Detector	FI

Compound	Retention time (min)
Quinoline[a]	2.2
Caprolactam	4.4

Note: Column packing — P1 = 10% Carbowax 20M® on Chromosorb W-AW-DMCS®.

[a] Internal standard.

Zillio-Grandi, F., Giovanni, M. S., and Callegaro, P., *Anal. Chem.,* 41, 1847, 1969.

Table GC 45
CAPROLACTAM MONOMER IN NYLON 6 WATER EXTRACTABLES

Column packing	P1
Temperature (°C)	195
Gas; flow rate (mℓ/min)	He; 130
Column	
Length, ft	6
Diameter, in. O.D.	1/8
Material	SS
Detector	FI

Compound	Retention time (min)
Caprolactam	0.7
Bis-[2-(2-methoxyethoxy)ethyl] ether (internal standard)	1.4

Note: Column packing — P1 = 10% silicone gum rubber SE-30 on Diatoport S® (60/80 mesh).

Ongemach, G. C. and Moody, A. C., *Anal. Chem.*, 39, 1005, 1967.

Table GC 46
DICHLOROMETHANE IN POLYCARBONATES

Direct Injection

Weigh 2 g polycarbonate into a septum sealed vial containing 10 mℓ $CHCl_3$. Heat at 50°C for 60 min to dissolve the polymer. Cool to room temperature and add 0.1 mℓ of the internal standard solution (1% tetrahydrofuran in $CHCl_3$ or dioxane). Inject a 3 $\mu\ell$ portion of the solution onto Column P1.

Headspace Analysis

Weigh 2 g of polycarbonate into a septum-sealed vial containing 9.9 mℓ of 1,4-dioxane. Heat at 60°C for 60 min to dissolve the polymer. Add 0.1 mℓ of the internal standard solution. Equilibrate at 60°C for 60 min. Inject a portion of the headspace gas onto Column P2.

Column packing	P1	P2
Temperature (°C)	60	70
Gas; flow rate (mℓ/min)	N_2; 30	N_2; 30
Column		
Length, ft	10	10
Diameter, in.	1/8	1/8
Material	SS	SS
Detector	FI	FI

Compound	Retention time (min)	
Dichloromethane	1.1	4.5
Tetrahydrofuran (THF)	1.7	7.0

Note: Column packing — P1 = 10% 2-Ethylhexyl Sebacate on Chromosorb W® (60/80 mesh).
P2 = 10% 2-Ethylhexyl Sebacate on Chromosorb W® (60/80 mesh).

DiPasquale, G., DiIorio, G., and Capaccioli, T., *J. Chromatogr.*, 152, 538, 1978.

Table GC 47
ANTIOXIDANTS AND ULTRAVIOLET ABSORBERS IN POLYETHYLENE

Place 1 g of sample and 15 mℓ of toluene in a 250-mℓ flask equipped with a reflux condenser and magnetic stirrer. Heat and stir until complete dispersion occurs. While stirring add 20 mℓ of ethanol or chloroform to precipitate the polymer. Filter to remove the polymer and evaporate the filtrate to dryness. Dissolve the residue in 5 mℓ of diethyl ether or ethanol. Chromatograph a 5-μℓ aliquot.

Column packing	P1	P1
Temperature (°C)	250	300
Gas; flow rate (mℓ/min)	Ar; 45	Ar; 45
Column		
Length, ft	2.5	2.5
Diameter, in.	1/4	1/4
Material	SS	SS
Detector	FI	FI

		Relative retention time	
Chemical name	**Trade name**	**P1**	**P2**
2,6-Di-*t*-butyl-*p*-cresol	Topanol C	0.05	
Tris(*p*-nonylphenyl)phosphite	Polygard	0.09	
Pentaerythritol tetra-3-(3,5-di-*t*-butyl-4-hydroxyphenyl)propionate	Irganox 1010	0.13	
2,2-Bis(3-methyl-4-hydroxyphenyl)propane	Nonox DCP	0.30	
1,3-Di(3-*t*-butyl-4-hydroxy-6-methylphenyl) butane	Anullex PBA15®	0.85	0.85
4,4′-Thiobis(6-*t*-butyl-*m*-cresol)	Santonox R®	1.00	1.00
Bis[2-hydroxy-5-methyl-3-(1-methylcyclohexyl)phenyl]methane	Nonox WSP®	2.63	2.63
Octadecyl-3-(3,5-di-*t*-butyl-4-hydroxyphenyl) propionate	Irganox 1076®		5.32
1,1,3-Tris(2-methyl-4-hydroxy-5-*t*-butylphenyl)butane	Topanol CA®		6.35
1,3,5-Trimethyl-2,4,6-tris(3,5-di-*t*-butyl-4-hydroxybenzyl)benzene	Ionox 330®		47.7
2-(2′-Hydroxy-5′-methylphenyl)benzotriazole	Tinuvin P®	0.20	
2-(2′-Hydroxy-3′-*t*-butyl-5′-methylphenyl)3-chlorobenzotriazole	Tinuvin 326®	0.64	
2-(2′-Hydroxy-3′,5′-di-*t*-butylphenyl)-3-chloro-benzotriazole	Tinuvin 326®	0.85	
2-Hydroxy-4-*n*-octoxygenzophenone			

Note: Column packing — P1 = 10% silicone gum rubber E301 on Diatomite CQ® (80/100 mesh).

Denning, J. A. and Marshall, J. A., *Analyst,* 97, 710, 1972.

Table GC 48
2,6-DI-*TERT*-BUTYL-4-METHYLPHENOL IN POLYETHYLENE

Weigh 10 g of polyethylene powder into a bottle and add 100 mℓ of hexane. Cap securely and shake the bottle to allow complete contact with the solvent. Allow to stand for 5 hr and chromatograph an aliquot of the supernatant liquid.

Column packing	P1
Temperature (°C)	200
Gas; flow rate (mℓ/min)	N_2; 70
Column	
Length, m	2
Diameter, in.	1/4
Material	Cu
Detector	FI

Compound	Retention time (min)
2,6-Di-*tert*-butyl-4-methylphenol	7
2,6-Di-*tert*-butyl-4-ethylphenol	9

Note: Column packing — P1 = 20% SE-30 on Chromosorb W-HMDS® (60/80 mesh).

Long, R. E. and Guvernator, G. C., *Anal. Chem.*, 39, 1493, 1967.

Table GC 49
DIETHYLENE GLYCOL IN POLY(ETHYLENE TEREPHTHALATE)

Weigh 1 g of fiber or polymer into a 5/8-inch o.d. thick-walled borosilicate glass tube sealed at one end. Add 3 mℓ of H_2O and seal the other end of the tube. Place the tube in a pressure reaction vessel and heat at 230°C for 4 hr. Allow the chamber to cool to room temperature. Add 0.5 mℓ of benzyl alcohol (internal standard) to the hydrolysate and centrifuge. Chromatograph a 15-$\mu\ell$ aliquot (thermal conductivity detector) or a 5-$\mu\ell$ (flame ionization detector).

Column packing	P1
Temperature (°C)	180
Gas; flow rate (mℓ/min)	He; 50
Column	
Length, ft	10
Diameter, in. I.D.	0.055
Material	SS
Detector	TC or FI

Compound	Retention time (min)
Benzyl alcohol	6.5
Diethylene glycol	8.5

Note: Column packing — P1 = 10% Carbowax 20M® on Chromosorb W-HMDS® (60/80 mesh) or on Gas Chrom Z (60/80 mesh).

Ponder, L. H., *Anal. Chem.*, 40, 229, 1968.

Table GC 50
METHYLESTER END GROUPS IN POLY(ETHYLENE TEREPHTHALATE)

Reflux 10 mℓ of 80% hydrazine hydrate, containing a small amount of ethanol as an internal standard, with 2 g polymer for 30 min. Cool to room temperature and allow the terephthalic dihydrazide to settle out. Reaction with hydrazine liberates methanol present in PET as methylester end groups.

Column packing	P1
Temperature (°C)	65
Gas; flow rate (mℓ/min)	N_2; 8
Column	
Length, m	1
Length, in. O.D.	1/4
Material	Glass
Detector	FI

Compound	Retention time (min)
Methanol	1.2
Ethanol	4.5
Hydrazine	6

Note: Column packing — P1 = Tenax GC® (60/80 mesh).

Vink, D. and van Wijk, R., *Z. Anal. Chem.*, 264, 293, 1973.

Table GC 51
ACIDS, ALCOHOLS, AND GLYCOLS IN POLYESTERS, SILYL DERIVATIVES

Weigh 1-4 g of sample into a 100-mℓ flask and add 50 mℓ of 1N KOH in 2-ethoxyethanol. Reflux for 10 min with constant stirring. Cool to room temperature and adjust the pH to 1 with concentrated HCl. Add 25 mℓ pyridine to dissolve the organic acids and centrifuge to remove the KCl. React 50 μℓ of hydrolyzed sample with 500 μℓ of *N,O*-bis(trimethylsilyl)trifluoroacetamide at room temperature for 10 min. Chromatograph 0.1 μℓ of the silylated hydrolysate.

Column	P1	P2	P3	P4
Temperature (°C)	127	120—8 min 4 min to 210	183	145
Gas; flow rate (mℓ/min)	N_2; 10	N_2; 20	N_2; 33	N_2; 10
Column				
Length, ft	10	6	6	6
Diameter, in.	1/8	1/4	1/8	1/4
Material	SS	SS	SS	Glass
Detector	FI	FI	FI	FI

Compound	Retention time (min)			
Ethylene glycol-TMS	7.5	9.5		
Dodecane[a]	17.5			
Diethylene glycol-TMS	19			
Nonyl alcohol[a]		14.5		
1,4-Butanediol-TMS		17.5		

Table GC 51[52] (continued)
ACIDS, ALCOHOLS, AND GLYCOLS IN POLYESTERS, SILYL DERIVATIVES

Compound	Retention time (min)			
cis-1,4-Cyclohexane-dimethanol-TMS		30		
trans-1,4-cyclohexane-dimethanol-TMS	32			
Heptadecane[a]			16.5	
Isophthalic acid-TMS			20	
Terephthalic acid-TMS			23.5	
Methanol				3.8
Ethanol				7.5
Propanol[a]			18.8	

Note: Column packing — P1 = 3% Versilube F-50® on Chromosorb G-HP® (100/120 mesh).
P2 = 10% Versilube F-50® on Chromosorb W-HP® (100/120 mesh).
P3 = 10% Versilube F-50® on Chromosorb W-HP® (100/120 mesh).
P4 = Chromosorb 102® (60/80 mesh).

[a] Internal standard.

Allen, B. J., Essea, G. M., Keller, K. P., and Kinder, H. D., *Anal. Chem.,* 49, 741, 1977.

Table GC 52
WATER IN POLYESTERS

Prepare silylation reagent from 4 mℓ Tri-Sal(hexamethyldisilazone and trimethylchlorosilane), 1 mℓ $CHCl_3$ and 10 mℓ dry pyridine. Weigh 0.5 to 1.0 g sample into a 2-mℓ dry serum vial and add 1 mℓ reagent. Seal immediately with a Teflon®-lined cap using a crimping tool. Place the vial in an oven at 105°C for 40 min. Cool to ambient temperature and chromatograph a 1-μℓ aliquot. Water reacts with the components of the reagent to form hexamethyldisiloxane.

Column packing	P1
Temperature (°C)	50
Gas; flow rate (mℓ/min)	He; 33
Column	
Length, m	4
Diameter, mm I.D.	2
Material	Glass
Detector	FI

Compound	Retention time (min)
Trimethylchlorosilane	2
Chloroform, internal standard	3.5
Hexamethyldisiloxane	4.7
Pyridine	7
Hexamethyldisilazone	12

Note: Column packing — P1 = 1% OV-1 on Chromosorb W®.

Hoffman, R., *Anal. Chem.,* 48, 455, 1976.

Table GC 53
POLYESTER RESINS

Solvent and residual monomer in commercial resins are removed by precipitation. Weigh 10 g of sample into a beaker and add 100 mℓ of petroleum ether. After vigorous agitation allow the mixture to separate. Decant the supernatant liquid. Dissolve the residue in acetone and repeat the precipitation. Dry the residue in a vacuum oven at 50°C. Reflux the dried residue for about 4 to 18 hr with 100 mℓ of absolute methanol containing 0.1 g sodium methoxide. Chromatograph 20 to 30 μℓ of the resulting solution of dimethyl esters of the carboxylic acids and glycols.

Column packing	P1	P1	P1	P1
Temperature (°C)	180	110—180 8°/min	150	110
Gas; flow rate (mℓ/min)	He; 8	He; 8	He; 8	He; 8
Column				
Length, ft	12	12	12	12
Diameter, in. O.D.	1/4	1/4	1/4	1/4
Material	Cu	Cu	Cu	Cu
Detector	TC	TC	TC	TC

Compound (Dimethylester)	**B.P., °C**	**Retention time (min)**			
Acids					
Fumaric acid	192	2.8			—
Maleic acid	205	2.8	—	—	—
Itaconic acid	208	3.8	—	—	—
Methoxysuccinic	208	4.7	11.0	—	—
Adipic acid	116;13 mm	7.4	13.9	—	—
Orthophthalic acid	284	18.8	21.2	—	—
Isophthalic acid	284	24.6	25.4	—	—
Glycols					
Propylene	189	0.8	2.7	1.2	3.0
Ethylene	197	0.8	2.7	0.9	2.2
1,3-Butylene	207	1.4	—	—	—
Neopentyl	—	1.9	—	—	—
Dipropylene	232	3.2	9.2	—	—
Dimethylene	245	2.3	7.4	—	—
Triethylene	290	7.4	—	—	—

Note: Column packing — P1 = 16.7% GE Silicone SF-96® on Fluoropak 80®.

Perceval, D. F., *Anal. Chem.*, 35, 236, 1963.

Table GC 54
POLYHYDRIC ALCOHOLS

Column packing	P1
Temperature (°C)	150
Gas; flow rate (mℓ/min)	He; 60
Column	
Length, m	1
Diameter, in. O.D.	1/4
Material	Cu
Detector	FI

Compound	Retention time (min)
Water[a]	1.5
2,3-Butanediol	7.2
1,2-Propanediol	9.1
Ethylene glycol	12.2
1,3-Butanediol	17.5
1,3-Propanediol	24.1
1,4-Butanediol	43.1

Note: Column packing — P1 = 10% THEED on Chromosorb W® (60/80 mesh).

[a] Solvent.

Swinehart, B. A., *Anal. Chem.*, 40, 427, 1968.

Table GC 55
POLYHYDRIC ALCOHOLS IN OILS AND RESINS, TMS DERIVATIVES

Weigh a sample containing about 2 g of nonvolatile material into a 125-mℓ flask. Add 3 mℓ of methylene chloride and evaporate the solvents at 60°C using a current of air. Add exactly 200 mg of 2,4-butanediol and about 10 mℓ of butylamine. Reflux under a condenser for 2 hr to release the polyols from the resin. Add 0.5 mℓ of water and reflux for an additional 2 hr.

Place 5 drops of the aminolysis mixture into a micro test tube, add 0.5 mℓ of TMS reagent (20 parts bis(trimethylsilyl)fluoroacetamide and 80 parts hexamethyldisilizane — store at 60°C in the absence of moisture) and heat for 30 min at 60°C. Chromatograph 40 μℓ of the resulting trimethylsilyl ether solution.

Column packing	P1
Temperature (°C)	100
Gas; flow rate (mℓ/min)	He; 80
Column	
Length, ft	16
Diameter, in.	1/4
Material	Cu
Detector	n.a.

Compound (Trimethylsilyl ether)	Relative retention time[a]
Ethylene glycol	0.58
Propylene glycol	0.63
2,3-Butanediol	0.73
1,3-Butanediol	0.83
Neopentyl glycol	0.86
1,4-Butanediol	1.00
Diethylene glycol	1.19
Glycerine	1.29
Trimethylol ethane	1.46
Trimethylol propane	1.63
Triethylene glycol	1.83
Pentaerythritol	1.88

Note: Column packing — P1 = 20% Silicone Grease DC-11® on Chromosorb W® (60/80 mesh).

[a] Relative to 1,4-Butanediol.

Esposito, G. G. and Swann, M. W., *Anal. Chem.*, 41, 1118, 1969.

Table GC 56
ADDITIVES IN POLYPROPYLENE

Add 3.0 g polymer to 150 mℓ *p*-xylene and heat at reflux until solution is complete. Add 1.0 mℓ of a 0.3% solution of biphenyl (internal standard) in *p*-xylene. Dilute with 15 mℓ of *p*-dioxane, cool in an ice bath and filter. Concentrate the filtrate to about 1 mℓ in a rotary evaporator at 45 to 50°C. Dilute to 6.0 mℓ with *p*-dioxane.

Column packing	P1
Temperature (°C)	100—340 15°/min
Gas; flow rate (mℓ/min)	He; n.a.
Column	
Length, ft	5
Diameter, in.	1/4
Detector	FI

Compound	Approximate retention time (min)
Biphenyl (internal standard)	6.8
2,6-Di-*tert*-butyl-*p*-cresol	7.8
4,4′-Methylenebis[2,6-di-*tert*-butylphenol]	Elutes within 1 hr
2,2′-Methylenebis[6-(1-methyl-cyclohexyl)-*p*-cresol]	Elutes within 1 hr
4,4′-Thiobis-6-*tert*-butyl-*m*-cresol	Elutes within 1 hr
4,4′-Butylidenebis[6-*tert*-butyl-*m*-cresol]	Elutes within 1 hr
1,3,5-Trimethyl-2,4,6-tris(3,5-di-*tert*-butyl-4-hydroxybenzyl)benzene	Elutes within 1 hr
Octadecyl-3,5-di-*tert*-butyl-4-hydroxy-hydrocinnamate	Elutes within 1 hr
Tris(2,6-di-*tert*-butyl-4-hydroxy-phenyl)phosphate	Elutes within 1 hr
2-Hydroxy-4-methoxybenzophenone	Elutes within 1 hr
2-Hydroxy-4-(octyloxy)benzophenone	Elutes within 1 hr
4(Dodecyloxy)-2-hydroxybenzophen-one	19.8
2-(2H-Benzotriazol-2-yl)-*p*-cresol	Elutes within 1 hr
2,4-Di-*tert*-butyl-6-(5-chloro-2H-benzotriazol-2-yl)phenol	Elutes within 1 hr
2,2′-Thiobis(6-*tert*-butyl-*p*-cresol)	Elutes within 1 hr
Dilauryl-3,3′-thiodipropionate	Elutes within 1 hr
Dioctadecyl-3,5-di-*tert*-butyl-4-hydroxy-benzylphosphonate	Elutes within 1 hr
Distearyl pentaerythritol diphosphate	Elutes within 1 hr

Note: Column packing — P1 = 10% SE-30 on Chromosorb W® (40/60 mesh).

Lappin, G. R. and Zannucci, J. S., *Anal. Chem.*, 41, 2076, 1969.

Table GC 57
POLYURETHANE POLYETHERS

Prepare reagent by dropwise addition of 80 g acetic anhydride to 120 g *p*-toluene sulfonic acid at room temperature. Reflux 30 min at 120°C. Polyethers are cleaved by reacting 100-mg polyether with 2 mℓ of reagent for 2 hr at 120°C. Neutralize the reaction mixture with 50% aq. NaOH and extract with 20 mℓ diethyl ether. Evaporate the extract to a small volume and inject a 10-μℓ aliquot.

Column packing	P1	P1	P2
Temperature (°C)	170	65	230
Gas; flow rate (mℓ/min)	He; 60	He; 60	He; 60
Column			
Length, m	1	1	1
Diameter, mm I.D.	3	3	3
Material	SS	SS	SS
Detector	TC	TC	TC

Compound	**Retention time (min)**		
Propylene glycol diacetate[a]	1		
Glycerol triacetate[a]	2		
Trimethylolpropane triacetate[a]	4		
Triethanolamine triacetate[a]	11.3		
Pentaerythritol tetraacetate[a]	13		
Propylene glycol diacetate[b]		10.3	
Ethylene glycol diacetate[b]		14	
Propylene glycol diacetate[c]			0.5
Derivative of 1,2-diaminoethane[c]			2.5
Derivative of 2,2-diaminodiethylamine[c]			10

Note: Column packing — P1 = 15% FFAP (free fatty acid phase) on Uniport B® (60/80 mesh).
P2 = 10% SE-30 silicone on Uniport B® (60/80 mesh).

[a] Polyurethane polyethers.
[b] Polyurethane polyethers based on 1,2-diaminoethane.
[c] Polyurethane polyethers based on amines.

Tsuji, K. and Konishi, K., *Analyst*, 99, 54, 1974.

Table GC 58
POLYOL BASE COMPOUNDS IN POLYURETHANE POLYETHERS

Prepare reagent by refluxing 126 g dry *p*-toluenesulfonic acid with 150 g acetate anhydride at 130°C for 30 min. Reflux 1 g sample and 5 g reagent at 130°C for 2.5 hr. Extract the reaction mixture with diethyl ether. Wash the ether solution with water to remove the excess reagent. Inject 1-$\mu\ell$ of the ether extract.

Column packing	P1
Temperature (°C)	180—260
	6°/min
Gas; flow rate (mℓ/min)	He; 30
Column	
Length, m	2
Diameter, mmI.D.	3
Material	SS
Detector	na

Compound	Elution time (°C)
Propylene glycol diacetate	188
Glycerol triacetate	205
Trimethylolpropane triacetate	225
Sorbitol acetate	250—260

Note: Column packing — P1 = 25% w/w Apiezon L® on Chromosorb W-AW® (60/80 mesh).

Tsuji, K. and Konishi,K., *Analyst,* 96, 457, 1971.

Table GC 59
RESIDUAL AROMATIC HYDROCARBONS IN POLYSTYRENE

Dissolve 1 g of polymer in 10 m of propylene oxide containing 0.1% (v/v) *n*-undecane. Gel, pigment and filler may remain as an undissolved suspension without detriment to the analysis. Chromatograph 5 $\mu\ell$ of the solution.

Column packing	Pl
Temperature (°C)	80
Gas; flow rate (mℓ/min)	He; 100
Column	
Length, cm	450
Diameter, cm	0.6
Material	Cu
Detector	FI

Compound	Relative retention time styrene = 1.00
Benzene	0.17
Toluene	0.29
n-Undecane[a]	0.40
Ethylbenzene	0.47
m-Xylene	0.50
p-Xylene	0.50
Cumene	0.62
o-Xylene	0.66
n-Propylbenzene	0.77
m-Ethyltoluene	0.84
p-Ethyltoluene	0.84
Isobutylbenzene	0.92
tert-Butylbenzene	0.92
sec-Butylbenzene	1.00
Styrene	1.00
o-Ethyltoluene	1.05
m-Diethylbenzene	1.35
p-Diethylbenzene	1.44
n-Butylbenzene	1.44
α-Methylstyrene	1.60
o-Methylstyrene	1.86
m-Methylstyrene	1.86
p-Methylstyrene	1.86

Note: Column packing — Pl = 10% Carbowax 15 to 20 M® on Celite® (60/72 mesh).

[a] Internal Standard.

Crompton, T. R. and Myero, L. W., *Eur. Polym. J.*, 4, 355, 1968.

Crompton, T. R., *Eur. Polym. J.*, 4, 473, 1968.

Table GC 60
2,6-DI-*TERT*-BUTYL-*P*-CRESOL AND 2(2′-HYDROXY-5′-METHYLPHENYL) BENZOTRIAZOLE IN POLYSTYRENE

Dissolve 5 g of polymer in 100 mℓ of CS_2. Chromatograph 2-μℓ aliquot for the determination of about 0.3% of the additives.

Column packing	Pl	Pl
Temperature (°C)	135	220
Gas; flow rate (mℓ/min)	N_2; 30	N_2; 43
Column		
Length, in.	12	12
Diameter, in. O.D.	1/8	1/8
Material	SS	SS
Detector	FI	FI

Compound	Retention time (min)	
2,6-di-*tert*-butyl-*p*-cresol	1.2	
2(2′-hydroxy-5′-methyl-phenyl)benzotriazole		4.4

Note: Column packing — Pl = 25% LAC-2R-446 plus 2% H_3PO_4 on Chromosorb W® (80/100 mesh).

Roberts, C. B. and Swank, J. D., *Anal. Chem.*, 36, 271, 1964.

Table GC 61
VINYL END GROUPS IN SILOXANE POLYMERS

Weigh 100 μℓ of diethyltetramethyldisiloxane (internal standard) and 10 g of sample into a 20-mℓ serum bottle. Seal with a silicone rubber cap and mix the contents overnight. Transfer 70 mg sample to a reactor and add 450 μℓ of 90% sulfuric acid. Heat from 65°C to 250°C at a rate of 10°C/min. Pass helium through the solution and condense the effluent in a liquid nitrogen trap. Chromatograph an aliquot of the trapped effluent and measure the ratio of ethane (formed from diethyltetramethyldisiloxane) to ethylene (liberated from vinyl end groups).

Column packing	Pl
Temperature (°C)	ambient
Gas; flow rate (mℓ/min)	He; 71
Column	
Length, ft	6
Diameter, in. O.D.	3/16
Material	SS
Detector	TC

Compound	Retention time (min)
Ethylene	3.5
Ethane	5.0

Note: Column packing — Pl = Porapak S® (80/100 mesh).

Bissell, E. R. and Fields, D. B., *J. Chromatogr. Sci.*, 10, 164, 1972.

Table GC 62
TRACE IMPURITIES IN STYRENE MONOMER

The 10% stationary phase solutions are forced through the capillary columns with Argon at 20 psig. Purging with Argon is continued until all of the solvent is evaporated from the column.

Column packing	P1	P2	P3
Temperature (°C)	100	65	95
Gas; psig	Ar; 19	Ar; 20	Ar; 40
Column			
Length, ft	200	100	200
Diameter, in. O.D.	.010	.006	.010
Material	SS	SS	SS
Detector	Triode Argon	Triode Argon	Triode Argon

		Relative retention time (Relative to benzene)		
Compound	**B.P., °C**	**I[a]**	**II[b]**	**III[c]**
Benzene	80.1	1.00	1.00	1.00
Toluene	110.6	1.13	1.37	1.41
Ethylbenzene	136.2	1.64	2.26	2.18
p-Xylene	138.3	1.75	2.39	2.22
m-Xylene	139.1	1.75	2.45	2.30
o-Xylene	144.4	1.93	2.94	2.72
Styrene	145.2	1.93	3.78	3.08
Cumene	152.5	2.04	3.02	2.94
n-Propylbenzene	159.45	2.35	3.67	
m-Ethyltoluene	161.3	2.48	4.00	3.80
p-Ethyltoluene	162.0	2.53	4.00	3.72
o-Ethyltoluene	165.1	2.77		
α-Methylstyrene	165.4	2.77	6.08	5.16
tert-Butylbenzene	168.8	2.75	4.43	4.34
Isobutylbenzene	171.0	2.99	4.69	4.46
sec-Butylbenzene	173.5	2.99	4.86	4.75
β-Methylstyrene	176.0	3.61		
Indane	177.0	3.87		
m-Cymene	175.2	3.15		
p-Cymene	177.1	3.31	5.47	5.21
o-Cymene	178.3	3.48		
m-Vinyltoluene	170.9	3.08		
p-Vinyltoluene	170.6	3.14		
o-Vinyltoluene	171.6	3.14		

Note: Column packing — P1 = Apiezon L®, 10% solution in methylene chloride.
P2 = Ucon 50-HB-2000,® 10% solution in absolute methanol.
P3 = Bis(phenoxyphenyl)ether, 10% solution in diethyl ether.

[a] Retention time in min = relative retention time × 9.2.
[b] Retention time in min = relative retention time × 4.9.
[c] Retention time in min = relative retention time × 9.1.

Hollis, O. L., *Anal. Chem.*, 33, 352, 1961.

Table GC 63
RESIDUAL MONOMERS IN STYRENE-ACRYLONITRILE COPOLYMER

The monomers are extracted from the copolymer with $CHCl_3$.

Column	20% polysorbate 80 on 60/80 Chromosorb P
Temperature (°C)	80
Gas; flow rate (mℓ/min)	Nitrogen; 30
Column	
Length, m	3
Diameter, mm	2
Material	Copper
Detector	FID

Chopra, S. K., Vishnoi, S. C., Kapoor, V. B., and Gupta, P. L., *Anal. Abstr.*, 36(6), 570, 1979.

Table GC 64
STYRENE IN THE ATMOSPHERE NEAR INDUSTRIAL SITES

Styrene was concentrated from air using an absorption flask containing acetic acid. Ethylbenzene was incorporated as an internal standard. Recoveries greater than 97% were obtained.

Column	10% FFAP on 80/100 Chromosorb W®
Temperature (°C)	120
Gas; flow rate (mℓ/min)	Nitrogen; 25
Column	
Length, m	2
Diameter, mm	2
Material	SS
Detector	FID

Choi, K. K. and Fung, K. W., *Analyst*, 104, 455, 1979.

Table GC 65
STYRENE OLIGOMERS

Column packing	P1
Temperature (°C)	170—260
	2.5°/min$_6$
Gas; flow rate (mℓ/min)	N_2; 1.5
Column	
Length, m	50
Diameter, mm	0.25
Detector	F1

Compound	Retention time (min)
Dimer 1	26.2
Dimer 2	27.8
Dimer 3	30.0
Dimer 4	31.1
Trimer 1	72.0
Trimer 2	85.1
Trimer 3	88.3
Trimer 4	92.1

Compounds and Structures

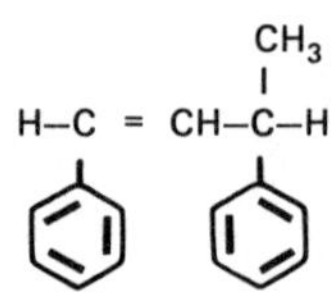

1,3-diphenylbutene-1

Dimer 2

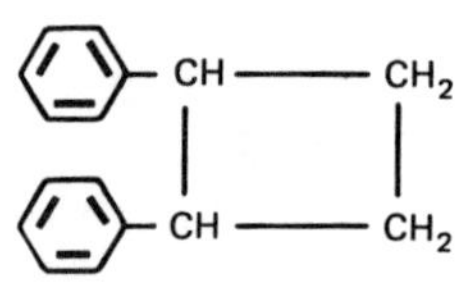

trans-1,2-diphenylcyclobutane

Dimer 3

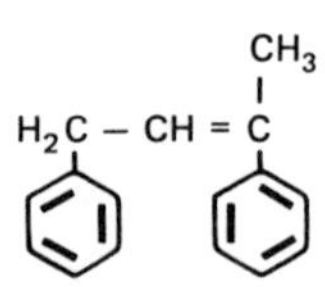

1,3-diphenylbutene-2

Dimer 4

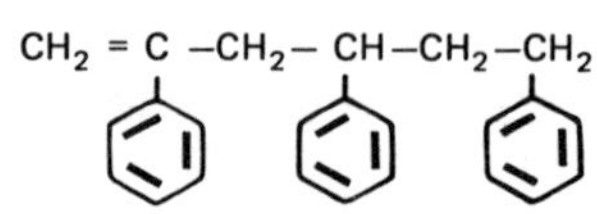

2,4,6-triphenylhexene-1

Trimer 1

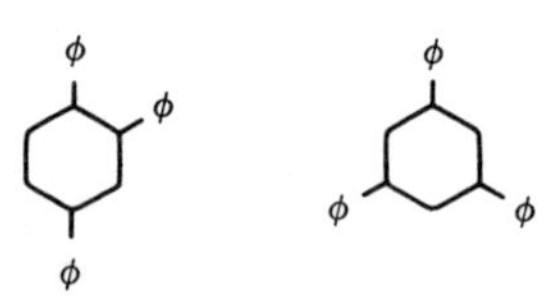

Isomers of triphenylcyclohexane and cis-trans configurations of 1,3,4-triphenylcyclohexane

Trimer 2, Trimer 3, Trimer 4

Note: Column packing — P1 = Hochstrakuumfett BASF® (unpolar, temperature fest.).

Stein, von D. J. and Mosthaf, H., *Angew. Makromol. Chem.*, 2, 39, 1968.

Table GC 66
VOLATILE COMPONENTS IN STYRENE POLYMERS

Volatile components are determined by automatic headspace analysis of polymers dissolved in dimethylformamide. Equilibration requires about 2 hr at 70 to 75°C.

Column packing	P1
Temperature (°C)	100
Gas; flow rate (mℓ/min)	N_2; 43
Column	
Length, m	2
Diameter, mm	3
Detector	FI

Compound	**Relative retention time Relative to n-butylbenzene**
Acrylonitrile	0.0324
Toluene	0.0434
Ethylbenzene	0.0852
m + *p*— Xylene	0.0967
Cumene	0.1662
o-Xylene	0.1944
n-Propylbenzene	0.2379
Styrene	0.3584
n-Butylbenzene	1.0000
α-Methyltoluene	1.0059

Note: Column packing — P1 = 15% Reoplex 400® on Embacel® (60/100 mesh).

Rohrschneider, L., *Z. Anal. Chem.*, 255, 345, 1971.

Table GC 67
VOLATILE COMPONENTS IN STYRENE AND VINYL CHLORIDE BASED POLYMERS

Dissolve or disperse the polymers in purified dimethylformamide. Inject an aliquot of the DMF solution onto the column. The polymer remains in the injection block and should be removed, after every 50 injections, by reaming. The method is designed for the determination of 0.01 to 1% volatile components in polymers.

Column packing	P1	P2
Temperature (°C)	120	120
Gas; flow rate (mℓ/min)	N_2; 30	N_2; 25
Column		
Length, ft	13	10
Diameter, in. O.D.	1/8	1/4
Detector	FI	FI

Compound	Relative retention time Relative to toluene	
Butadiene	0.40	
Acrylonitrile	0.76	
Toluene	1.00	1.00
Ethylbenzene	1.40	
Styrene	2.25	
Vinyl chloride		0.33
Vinyl acetate		0.58

Note: Column packing — P1 = Two columns in series: 3 ft column — 20% Tween 81® on Chromosorb W® (30/60 mesh), 10 ft column — 10% Resoflex 446® on Chromosorb W® (30/60 mesh).
P2 = 10% Resoflex 446® on Chromosorb W® (30/60 mesh).

Shapras, P. and Claver, G. C., *Anal. Chem.* 36, 2282, 1964.

Table GC 68
TRACE ANALYSIS OF VINYL CHLORIDE IN AIR

A selective GC column for the determination of vinyl chloride in air has been developed. The selectivity was tested with 21 possible pollutants. All of the tested compounds were separated from vinyl chloride. Relative retention times are shown below.

Column	mixture of Porapak S® and T (80:20)
Temperature (°C)	100
Gas; flow rate (mℓ/min)	Helium; 50
Column	
Length, m	1.5
Diameter, mm	4
Material	SS
Detector	FID

Formula	Component	Relative retention time Relative to vinyl chloride
CH_4	Methane	0.06
C_2H_4	Ethylene	0.15
C_2H_6	Ethane	0.17
C_2H_2	Acetylene	0.17
CHF_2Cl	Difluorochloromethane	0.28
C_3H_8	Propane	0.43
C_3H_6	Propene	0.43
C_3H_4	Propadiene	0.54
CF_2Cl_2	Difluorodichloromethane	0.57
CH_3Cl	Chloromethane	0.60
C_3H_4	Methylacetylene	0.60
MeOH	Methanol	0.66
CH_2O	Formaldehyde	0.68
H_2O	Water	0.78
C_2H_3Cl	Vinyl chloride	1.00
n-C_5H_{12}	Neopentane	1.11
i-C_4H_{10}	Isobutane	1.18
C_2H_4O	Ethylene oxide	1.18
CH_3CHO	Acetaldehyde	1.25
1,3-C_4H_6	1,3-Butadiene	1.50
i-C_5H_{12}	Isopentane	1.63
C_2H_5Cl	Chloroethane	1.90

Krockenberger, D., Lockewski, H., and Rohrschneider, L., *Chromatographia,* 12(12), 787, 1979.

Table GC 69
VINYL CHLORIDE MONOMER IN THE ENVIRONMENT

Vinyl chloride is trapped on C_{22} firebrick followed by polyethylene glycol 400 on Chromosorb W cooled in solid carbon dioxide in methanol. The vinyl chloride is desorbed and analyzed by gas chromatography.

Column	Ucon LB-550® on Chromosorb W®
Temperature (°C)	50
Gas; flow rate (mℓ/min)	Nitrogen
Column	
Length	n.a.
Diameter	n.a.
Material	n.a.
Detector	FID

Tanaka, K., Fukaya, K., Nishiyaina, N., Wada, Y., and Fukuli, S., *Gas Chromatogr. Abstr.*, 22, 147, 1979.

Table GC 70
VINYL CHLORIDE MONOMER IN POLYMERS AND EXTRACTS

Vinyl chloride monomer in *n*-heptane, water, and ethanol extracts of food products are determined by direct injection of the solution or extract. For the determination of residual vinyl chloride monomer in plastics, dissolve 100 mg of the sample in 8 mℓ of *N*,N′-dimethylacetamide.

Column packing	P1
Temperature (°C)	50/6 min isothermal 6 min 16°/min to 200°
Gas; flow rate (mℓ/min)	He; 25
Column	
Length, ft	25
Diameter, in. O.D.	1/8
Material	SS
Detector	FI

Compound	Retention time (min)
Vinyl chloride	4.7

Note: Column packing — P1 = 30% SE-52 on Chromosorb W-AW® (80/100 mesh).

Pu, H. H. and Hicks, S. Z., Jr., *J. Chromatogr.*, 132, 495, 1977.

Table GC 71
RESIDUAL VINYL CHLORIDE MONOMER IN POLY(VINYL CHLORIDE)

Dissolve 0.2 g polymer in 10 mℓ 1,2-dichloroethane (EDC) by warming to 40°C. Chromatograph a 1 μℓ aliquot. Taking the minimum detectable response as a signal-to-noise ratio of two, the detection limit is 0.03 μg VCM/mℓ EDC which is equivalent to 1.5 ppm VCM in PVC.

Column packing	P1
Temperature (°C)	90°
Gas; flow rate (mℓ/min)	Ar; 25
Column	
Length, m	5.5
Diameter, in.	1/8
Material	Al
Detector	FI

Compound	Relative retention time Relative to 1,2-dichloroethane
Vinyl Chloride Monomer (VCM)	0.32
1,2-Dichloroethane (EDC)	1.0

Note: Column packing — P1 = 3% Carbowax 20M® on Chomosorb G-AW-DMCS® (80/100 mesh).

Ravey, M. and Klopstock, J., *J. Chromatogr. Sci.,* 13, 552, 1975.

Table GC 72
TRACE IMPURITIES IN VINYL CHLORIDE MONOMER

Column packing	P1	P2
Temperature (°C)	40—150	50—150
	6°/min	6°/min
Gas; flow rate (mℓ/min)	He; 60	He; 60
Column		
Length, m	6	2
Diameter, mm	3.175	4.76
Material	SS	SS
Detector	FI	FI

Program starts as soon as vinyl chloride elutes

Compound	Relative retention time	
	P1	P2
Methane		0.051
Ethylene		0.149
Ethylene/ethane	0.222	
Ethane		0.196
Propane	0.267	0.690
Propylene	0.305	0.641
Isobutane	0.334	
Acetylene	0.412	
Butane	0.434	
Butene-1	0.495	
Allene	0.542	
Allene/methyl chloride		0.755
trans-Butene	0.571	
cis-Butene-2	0.652	
1,2-Butadiene	0.770	
Methyl acetylene	0.817	
Methl chloride	0.934	
Vinyl chloride	1.00	1.00
1,3-Butadiene		1.208
trans-Butene-2		1.238
Ethyl chloride	1.385	
Ethyl chloride/*cis*-Butene-2		1.266
Acetaldehyde	1.559	
Benzene	2.183	
Ethylene dichloride	2.693	2.395

Note: Column packing — P1 = 25% Igepal CO-880® and Antarox BL-344® (75:25) on firebrick (60/80 mesh).
P2 = Porapak S® (80/100 mesh).

American Society for Testing and Materials, Recommended practice for the determination of the purity of vinyl chloride monomer, in *Annual Book of ASTM Standards, Part 35,* American Society for Testing and Materials, Philadelphia, Part 35, 1979, D-3834.

Table GC 73
IDENTIFICATION AND ANALYSIS OF PLASTICIZERS IN PVC

A quick and safe identification method is shown for plasticizers in PVC using programmed temperature gas chromatography. Plasticizers were identified by their retention index values measured by using normal parafins. Several stationary phases with quite different polarity are proposed with silicon polymers (SE-30, OV-17 and QF-1) and a porous polymer such as Tenax.® Thermodynamic values for the solution of plasticizers were measured by using isothermal chromatography with all of the stationary phases. Quantitative analysis of plasticizers was obtained with good accuracy after a previous extraction and the addition of normal parafin as internal standard.

Column packing	P1	P2	P3	P4
Temperature Program (°C)				
Initial	100	100	100	180
Final	340	340	228	360
Rate (°C/min)	10	10	10	10
Gas; flow rate (mℓ/min)		Nitrogen, n.a.		
Column				
Length, m	2	2	2	2
Diameter, mm	3	3	3	3
Material	SS	SS	SS	SS

		P1		P2		P3		P4	
Compound	**Symbol**	T_r,°C	I_r	T_r,°C	I_r	T_r,°C	I_r	T_r,°C	I_r
Trimethyl phosphate	TMPh	107	854	124	1110				
Triethyl phosphate	TEPh	125	1091	143	1285				
Dimethyl adipate	DMA	140	1205	160.5	1426			200	1384
Dimethyl phthalate	DMP	165	1425	189	1705	172	1971	234	1842
Diethyl phthalate	DEP	181.5	1558	205.5	1884	183	2110	240.5	1928
Dimethyl sebacate	DMS	186	1619	205	1863	184	2106	239.5	1884
Tributyl phosphate	TBPh	186.5	1623	204	1865	193.5	2195	237	1865
Diallyl phthalate	DAP	195	1708	215	2022	194	2242	255	2091
Dibutyl adipate	DBA	199	1731	206.5	1942	195	2225	244.5	1955
Diethyl sebacate	DES	200	1759	211.5	1962	193	2221	246.5	1998
Dibutyl phthalate	DBP	218	1930	232.5	2201			265	2312
Bis(methyleneglycol) phthalate	BMGP	219.5	1954	244.5	2352			282	2481
Triphenyl phosphate	TPPh	258	2359	283	2841			Remains in column	
Dioctyl adipate	DOA	260.5	2382	261	2543			285	2526
Bis(butyleneglycol) phthalate	BBGP	265	2438	281	2856			319	2985
Dicyclohexyl phthalate	DCHP	268.5	2472	291	2919			330	3126
Dioctyl phthalate	DOP	272.5	2519	280	2807			299	2732
Dioctyl sebacate	DOS	294	2778	297	2995				

Note: Column Packing — P1 = 1.5% SE-30 on 60/80 mesh Chromosorb Q® (60/80 mesh).
P2 = 2.5% OV-17 on 60/80 mesh Chromosorb Q® (60/80 mesh).
P3 = 2.5% OF-1 on 60/80 mesh Chromosorb Q® (60/80 mesh).
P4 = Tenax.®

Messadi, D. and Vergnaud, J.-M., *J. Appl. Polym. Sci.,* 24, 1215, 1979.

Table GC 74
PLASTICIZERS IN POLY(VINYL CHLORIDE)

Dissolve 0.2 g of poly(vinyl chloride) plastic in 10 mℓ tetrahydrofuran containing 0.2% (w/v) dibutyl phthalate. Warm gently to facilitate dissolution. Inject 3.5 $\mu\ell$ of the solution.

Column packing	P1
Temperature (°C)	200
Gas; flow rate (mℓ/min)	N_2; 30
Column	
Length, m	1
Diameter, mm I.D.	2.6
Material	SS
Detector	FI

Compound	Retention time (min)
Tetrahydrofuran	0.1
Texanol isobutyrate[a]	0.83
Dibutyl phthalate	4.13
Butylbenzyl phthlate	5.7

Note: Column packing — P1 = 10% SE-30 on Chromosorb W HP® (80/100 mesh).

[a] 1,3-diisobutyrate-2,2,4-trimethylpentane.

Hayes, W. P. and Steel, P., *J. Chromatogr.*, 139, 395, 1977.

Table GC 75
1,1,1-TRICHLOROETHANE IN VINYL CHLORIDE POLYMERS AND FOOD

1,1,1-Trichloroethane was determined in poly(vinyl chloride) bottles used in food storage by analysis of the headspace vapor over an equilibrium mixture of H_2O and dimethylforamide solution of the sample after incubation at 70°C for 2 hr in a sealed vial.

Column	20% Apiezon L® on Diatomite C®
Temperature (°C)	—
Column	
Length, m	3
Diameter, in.	0.125
Material	SS
Detector	FI

Gilbert, J., Startin, J. R., and Wallwork, M. A., *Anal. Abstr.*, 36(6), 629, 1979.

IV.2 PYROLYSIS — GAS CHROMATOGRAPHY TABLES

INTRODUCTION

For the most part, tables are arranged in alphabetical order by polymer. Some non-polymer applications of this technique are listed at the end of this Section. The reader is referred to the list of tables in Section III or the master index for specific compounds.

Please note n.a. = not available from the original source.

Table P-GC 1
ACRYLONITRILE POLYMER AND COPOLYMER

Sample	S1		S2
Pyrolyzer type	F		F
Pyrolysis temperature (°C)	470		470
Pyrolysis time (sec)	n.a.		n.a.
Sample size pyrolyzed (mg)	0.5		0.5
Temperature		60 — 200 @ 16°/min	
Gas; flow rate (mℓ/min)	N_2		N_2
Column			
Length, m	2		2
Diameter, mm I.D.	3		3
Material	n.a.		n.a.
Detector	MS		MS

Compound	Approximate retention time (min)	
$CH_2=CHCOOCH_3$		2.8
CH_3OH	4	3.3
$CH_2=CHCN$		4
$H_3COOCCH=CHCH_2COOCH_3$		12.1
$H_3COOCCH=CHCHCH_3$ (CH bearing $COOCH_3$)		13
$NCCH=CHCH_2COOCH_3$		13.8
$NCCH=CHCH_2CHCH_3$ (CH bearing $COOCH_3$)		14.2
$NCCH=CHCH_2CN$	15.5	15.5
$NCCH=CHCHCH_3$ / CN	17	17
$CH_2=CHCH_2CHCH_3$ / CN CN	19	19

Note: Sample — S1 = Polyacrylonitrile.
S2 = Acrylonitrile-methyl acrylate copolymer.
Pyrolyzer type — F = Furnace.
Column packing — 10% PEG-6000 on Diasolid L® (80/100 mesh).

Yamamoto, Y., Tsuge, S., and Takeuchi, T., *Macromolecules,* 5, 325, 1972.

Table P-GC 2
ACRYLONITRILE—CHLOROSTYRENE COPOLYMERS

Sample	S1	S2	S3
Pyrolyzer type	F	F	F
Pyrolysis temperature (°C)	530	530	530
Pyrolysis time (sec)	n.a.	n.a.	n.a.
Sample size pyrolyzed (mg)	300—500	300—500	300—500
Temperature (°C)		100—290° @ 10°/min	
Gas; flow rate (mℓ/min)	N_2; n.a.	N_2; n.a.	N_2; n.a.
Column			
Length, m	1	1	1
Diameter, mm I.D.	3	3	3
Material	SS	SS	SS
Detector	FI	FI	FI

Compound	Approximate retention time (min)		
Acrylonitrile	1		1
$CH_2=CHøCl$		4	4
CH_2-CH=CH CN CN	8		8
CH=CH-CH-CH_2 CN CN	9		9
CH_2-CH-CH_2-CH=CH_2 CN CN	11		11

Compound	Approximate retention time (min)		
CH=CH-CH_2øCl CN			13
CH_3-CH-CH_2-C=CH_2 CN øCl			16
CH=CH-CH_2 øCl øCl		17	17
CH=CHCH-CH_3 øCl øCl		19	19
CH_3-CH-CH_2-C=CH_2 øCl øCl		20	20

Note: Sample — S1 = Polyacrylonitrile.
S2 = Poly (chlorostyrene), meta or para.
S3 = Acrylonitrile — chlorostyrene (m,p) copolymer.
Pyrolyzer Type — F = Furnace.
Column Packing — 10% FFAP on Uniport B® (80/100 mesh).

Okumoto, T., Tsuge, S., Yamamoto, Y., Takeuchi, T., *Macromolecules,* 7, 376, 1974.

Table P-GC 3
CELLULOSE AND DERIVATIVES

Sample	S1	S2	S3
Pyrolyzer type	CP	CP	CP
Pyrolysis temperature (°C)	460	460	460
Pyrolysis time (sec)	5	5	5
Sample size pyrolyzed (mg)	0.2—0.4	0.2—0.4	0.2—0.4
Temperature (°C)		80—230° @ 7.5°/min	
Gas; flow rate (mℓ/min)	He; 40	He; 40	He; 40
Column			
Length, m	3	3	3
Diameter, mm I.D.	3	3	3
Material	SS	SS	SS
Detector	FI	FI	FI

Compound	Approximate retention time (min)		
2-Furaldehyde	13.5	13.5	
1,6-Anhydro-3,4-dideoxy-β-D-glucopyranosen-2-one	23.5		23.5
Levulinic acid	28		28
1,6-Anhydro-3-deoxy-β-D-glucopyroanosen	30		30
5-Hydroxymethyl-2-furaldehyde	31	31	31

Note: Sample — S1 = Microcrystalline cellulose.
S2 = D-glucose.
S3 = Levoglucosan.
Pyrolyzer type — CP = Curie Point.
Column packing — P1 = 5% Carbowax 20M® on Chromosorb GAW® (60/80 mesh).

Ohnishi, A., Kato, K., and Takagi, E., *Polym. J.*, 7, 431, 1975.

Table P-GC 4
CELLULOSE ESTERS

Sample	S1	S2	S3
Pyrolyzer type	PC	PC	PC
Pyrolysis temperature (°C)	700	700	700
Pyrolysis time (sec)	n.a.	n.a.	n.a.
Sample size pyrolyzed (mg)	1	1	1
Temperature (°C)	150	150	150
Gas; flow rate (mℓ/min)	He; 45	He; 45	He; 45
Column			
Length, ft	8	8	8
Diameter, in O.D.	0.25	0.25	0.25
Material	n.a.	n.a.	n.a.
Detector	n.a.	n.a.	n.a.

Compound	Approximate retention time (min)		
Acetic acid	4.5	4.5	4.5
Propionic acid		8.0	8.0
Butyric acid			15.5

Note: Sample — S1 = Cellulose acetate.
S2 = Cellulose propionate.
S3 = Cellulose butyrate.
Pyrolyzer type — PC = Platinum coil.
Column packing — P1 = 5% UCON® on Halopart F®.

Groten, B., *Anal. Chem.*, 36, 1206, 1964.

Table P-GC 5
CELLULOSE ESTERS AND ETHERS

Sample	S1	S2	S3	S4	S5
Pyrolyzer type	F	F	F	F	F
Pyrolysis temperature (°C)	400	400	400	360	360
Pyrolysis time (min)	7	7	7	7	7
Sample size pyrolyzed (mg)	2	2	2	2	2
Temperature (°C)	150	150	50	160	220
Gas; flow rate (mℓ/min)	n.a.	n.a.	n.a.	n.a.	n.a.
Column					
Length, ft	2	2	2	2	2
Diameter, in. I.D.	n.a.	n.a.	n.a.	n.a.	n.a.
Material	n.a.	n.a.	n.a.	n.a.	n.a.
Detector	n.a.	n.a.	n.a.	n.a.	n.a.

Compound	Approximate retention time (min)				
2-Ethylhexanoic acid	1				
3,5,5-Trimethylhexanoic acid		1.2			
Toluene			1.8		
Diphenylmethane				2.1	
Benzophenone				5.5	0.8
Triphenylmethane					2.8

Note: Sample — S1 = Cellulose 2-ethylhexanoate.
S2 = Cellulose 3,5,5-trimethylhexanoate.
S3 = Benzyl cellulose.
S4 = Benzohydryl cellulose.
S5 = Trityl cellulose.
Pyrolyzer type — F = Furnace.
Column packing — P1 = 10% silicone gum rubber SE-30 on Chromosorb P®.

King, W. D. and Stanonis, D. J., *TAPPI,* 52, 465, 1969.

Table P-GC 6
CHLORINATED POLYETHYLENE

Sample	S1	S2	S3
Pyrolyzer type	F	F	F
Pyrolysis temperature (°C)	660	660	660
Pyrolysis time (sec)	60	60	60
Sample size pyrolyzed (mg)	0.1—0.3	0.1—0.3	0.1—0.3
Temperature (°C)		50°—210° @ 16°/min	
Gas; flow rate (mℓ/min)	n.a.	n.a.	n.a.
Column			
Length, m	2	2	2
Diameter, mm I.D.	3	3	3
Material	n.a.	n.a.	n.a.
Detector	FI	FI	FI
Compound		**Approximate retention time (min)**	
Benzene	3.5	3.5	3.5
Toluene	4.8	4.8	4.8
Chlorobenzene			6.6
Styrene	7.0	7.0	
1,3-Dichlorobenzene			8.5
1,3,5-Trichlorobenzene			9.5
Naphthalene		12.5	

Note: Sample — S1 = Chlorinated polyethylene.
S2 = Polyvinyl chloride.
S3 = Polyvinylidene chloride.
Pyrolyzer type — F = Furnace.
Column packing — P1 = 10% Carbowax 6000® on Diasolid H® (80/100 mesh).
P2 = Precolumn of 25% NaOH on Diasolid H® (30—60 mesh).

Tsuge, S., Okumoto, T., and Takeuchi, T., *Macromolecules,* 2, 200, 1969.

Table P-GC 7
COUMARONE — INDENE RESINS

Pyrolyzer type	CP
Pyrolysis temperature (°C)	770
Time (sec)	10
Sample size pyrolyzed (mg)	n.a.
Temperature °C)	65—160° @ 4°/min
Gas; flow rate (mℓ/min)	N_2;23
Column	
Length, ft	6
Diameter, in. I.D.	1/8
Material	SS
Detector	FI

Compound	Retention time (min)
Cyclopentadiene	1.06
Benzene	1.57
Toluene	2.28
o-Xylene	3.50
Styrene	4.60
α-Methylstyrene	6.34
Coumarone	7.05
Indene	9.13
Methyl indene	12.13

Note: Pyrolyzer type — CP = Curie Point.
Column packing — 5% XE60 on Gas Chrom Q® (80/100 mesh).

Luke, B. G., *J. Chromatogr. Sci.*, 11, 435, 1973.

Table P-GC 8
EPOXY RESIN

Packing	P1	P2	P1
Sample	S1	S2	S2
Pyrolyzer type	NF	NF	NF
Pyrolysis temperature (°C)	700	700	700
Pyrolysis time (sec)	n.a.	n.a.	n.a.
Sample sie pyrolyzed (mg)	5—50	5—50	5—50
Temperature (°C)	50	50	50
Gas; flow rate (mℓ/min)	N_2; 40	N_2; 40	N_2; 40
Column			
Length, mm I.D.	n.a.	n.a.	n.a.
Diameter, mm I.D.	n.a.	n.a.	n.a.
Material	n.a.	n.a.	n.a.
Detector	C	C	C

Compound	Approximate retention time (min)		
Formaldehyde	1		1
Acetaldehyde	2.5		2.5
Acetone	8		8
Water	10		10
Benzene	18	1	18
Toluene	32	2	32
Phenol		5	
o-Cresol		7	

Note: Packing — P1 = 7,8-Benzoquinoline on Chromosorb P® (15/85 mesh).
P2 = Apiezon L® on Chromosorb P® (10/90 mesh).
Sample — S1 = EP274 epoxy resin hardened with 4,4′-diaminodiphenyl methane.
S2 = EP274 epoxy resin hardened with phthalic anhydride.
Pyrolyzer type NF = Heated nichrome filament.

Bishop, D. P. and Smith, D. A., *J. Appl. Polym. Sci.*, 14, 205, 1970.

Table P-GC 9
ETHYL ACRYLATE-METHYL METHACRYLATE COPOLYMER

Sample	S1	S2
Pyrolyzer type	n.a.	n.a.
Pyrolysis temperature (°C)	450	450
Pyrolysis time (sec)	60	60
Sample size pyrolyzed (mg)	1—2	1—2
Temperature (°C)	80°/6 min — 200° @ 10°/min	
Gas; flow rate (mℓ/min)	He; 30	He; 30
Column		
Length, ft	15	15
Diameter, in. I.D.	0.125	0.125
Material	SS	SS
Detector	TC	TC

Compound	Approximate retention time (min)	
CO_2 + olefin	1	1
Methanol	6	
Ethanol	7	7
Ethyl acrylate (EA)	9	
Methyl methacrylate (MMA)	9.5	9.5
Ethyl methacrylate (EMA)	10.5	10.5
Acrylic acid	20	20
Methacrylic acid	21	21

Note: Sample — S1 = Random 50/50 EA/MMA copolymer.
S2 = Block 50/50 EA/MMA copolymer.
Column packing — P = 20% Carbowax 20M-TPA® on Chromosorb W® (60/80 mesh).

Wallisch, K. L., *J. Appl. Polym. Sci.*, 18, 203, 1974.

Table P-GC 10[87]
GLYCIDYL METHACRYLATE — STYRENE COPOLYMER

Pyrolyzer type	CP[a]
Pyrolysis temperature (°C)	610
Pyrolysis time (sec)	10
Sample size pyrolyzed (mg)	n.a.
Temperature (°C)	60°—275° @ 5°/min
Gas; flow rate (mℓ/min)	N_2; 20
Column	
Length, m	4
Diameter, mm I.D.	3
Material	n.a.
Detector	FI

Compound	Approximate retention time (min)
Styrene monomer (S)	10
Glycidyl methacrylate monomer (G)	14
GS Dimer (mol wt 185)	22.5
GG Dimer (mol wt 258)	24
GS Dimer (mol wt 248)	26.5
SS Dimer (mol wt 208)	27.5
GG Dimer (mol wt 268)	28

Note: Column Packing — PI = 3% E-30 on Diatomite C′Q′® (100/200 mesh).

[a] CP = Curie Point.

Kalal, J., Zachoval, J., Kubat, J., and Svec, F., *J. Anal. Appl. Pyrol.*, 1, 143, 1979.

Table P-GC 11
HEXAFLUOROPROPYLENE-VINYLIDENE FLUORIDE COPOLYMERS

Pyrolyzer type	CP[a]
Pyrolysis temperature (°C)	800
Pyrolysis time (sec)	4
Sample size pyrolyzed (mg)	Dip; 10% acetone solution
Temperature (°C)	90°/3 min — 120°
Gas; flow rate (mℓ/min)	Ar; 15
Column	
Length, ft	10
Diameter, in. O.D.	1/8
Material	SS
Detector	FI

Compound	Approximate relative retention time (min)
Vinylidene fluoride	1.0
Trifluoromethane	1.5
Hexafluoropropylene	3.0

Note: Column Packing — P1 = β,β′-oxydipropionitrile Porasil C Durapak® (80/100 mesh).

[a] CP = Curie Point.

Blackwell, J. T., *Anal. Chem.*, 43, 1883, 1976.

Table P-GC 12
ION EXCHANGE RESINS

Sample	S1	S2	S3
Pyrolyzer type	CP	CP	CP
Pyrolysis temperature (°C)	980	980	980
Pyrolysis time (sec)	10	10	10
Sample size pyrolyzed (mg)	n.a.	n.a.	n.a.
Temperature (°C)	100	100	100
Gas; flow rate (mℓ/min)	N_2; 60	N_2; 60	N_2; 60
Column			
Length, cm	125	125	125
Diameter, mm I.D.	n.a.	n.a.	n.a.
Material	n.a.	n.a.	n.a.
Detector	FI	FI	FI

Product	Approximate retention time (min)		
Methyl chloride			0.3
Lights (C_2H_4, C_2H_2)	0.3	0.3	0.3
Benzene	1.1	1.1	1.1
Toluene	2.2	2.2	2.2
Ethyl benzene, xylene	4	4	4
Styrene	6.1	6.1	6.1
Ethyl toluene		7.7	7.7
Vinyl toluene		12.2	12.2

Note: Sample — S1 = Zeokarb 225® (acrylic, weak cation exchange).
S2 = Amberlite IRA 400® (cation exchange).
S3 = De-Acidite H-IP® (anion exchange).
Pyrolyzer type — CP = Curie Point.
Column packing — 10% Polyphenyl ether (5 rings) on Celite.®

Parrish, J. R., *Anal. Chem.*, 45, 1659, 1973.

Table P-GC 13
ION EXCHANGE RESIN

Sample	S1	S2
Pyrolyzer type	F	F
Pyrolysis temperature (°C)	800	800
Pyrolysis time (sec)	10	10
Sample size pyrolyzed (mg)	0.5	0.5
Temperature (°C)	100	100
Gas; flow rate (mℓ/min)	n.a.	n.a.
Column		
Length, m	1.8	1.8
Diameter, mm I.D.	5	5
Material	n.a.	n.a.
Detector	FI	FI

Compound	Approximate retention time (min)	
Ethyl benzene	14	14
p-Xylene		16
m-Xylene	18	18
o-Xylene		21

Note: Sample — S1 = Cation exchange (10% crosslinking).
S2 = Anion exchange.
Pyrolyzer type — F = Filament.
Column packing — 5% Bentone 34® plus 5% dinonylphthalate on Gas Chrom P® (70/80 mesh).

Parrish, J. R., *Anal. Chem.*, 47, 1999, 1975.

Table P-GC 14
ION EXCHANGE RESINS (ANIONIC)

Sample	S1	S2	S3
Pyrolyzer type	CP	CP	CP
Pyrolysis temperature (°C)	800	800	800
Pyrolysis time (sec)	2	2	2
Sample size pyrolyzed (μg)	300	300	300
Temperature (°C)		40°/5 min — 180° @ 4°/min	
Gas; flow rate (mℓ/min)	n.a.	n.a.	n.a.
Column			
Length, m	100	100	100
Diameter, mm I.D.	0.8	0.8	0.8
Material	SS	SS	SS
Detector	FI	FI	FI

Compound	Approximate retention time (min)		
Methylamine	1.8		1.8
Dimethylamine			1.8
Formaldehyde		2.0	
Acetaldehyde		2.0	
Methyl chloride	2.1	2.1	2.1
Trimethylamine		2.3	2.3
Ethyl chloride			2.4
Benzene	3	3	
Toluene	4.8	4.8	4.8
m,p-xylene	8.5	8.5	8.5
Styrene	9.5		9.5
m,p-Methylstyrene	16		
Phenol		17	
o-Cresol		19.5	
m-Ethylvinylbenzene	19.5		19.5
p-Ethylvinylbenzene	20		20
m-Divinylbenzene	21		21
Methylindole	24		
Dimethylindole	27		

Note: Sample — S1 = Dowex® 1 × 4 (Cl^-); STY/DVB matrix; Functional groups — $N^+ (CH_3)_3Cl^-$.
S2 = Amberlite IR 413® (Cl^-); Bisphenol, formaldehyde, triethylene-tetramine matrix; Functional groups — NH $(CH_2CH_2NH)_3$-H.
S3 = Amberlite IRA® 68(Cl^-); Acryl/DVB Matrix; Functional group — NH $(CH_2)_3$-$N(CH_3)_2$.
Pyrolyzer type — CP = Curie Point.
Column packing — P1 = Capillary SE30 coating.

Blasius, E. and Häusler, H., *Z. Anal. Chem.*, 276, 11, 1975.

Table P-GC 15
ION EXCHANGE RESIN (CATIONIC)

Column packing	P1
Sample	S1
Pyrolyzer type	CP
Pyrolysis temperature (°C)	770
Pyrolysis time (sec)	2
Sample size pyrolyzed (mg)	0.2—0.6
Column temperature (°C)	65°/15 min—180° @ 4°/min
Gas; flow rate (mℓ/min)	He; 30
Column	
Length, m	100
Diameter, mm I.D.	0.8
Material	Metal
Detector	FI

Compound	**Approximate retention time (min)**
Benzene	4
Toluene	6.5
Ethylbenzene	10
m- and *p*-Xylene	10.5
Styrene	12
Isopropylbenzene	15
Allyl benzene	17
n-Propylbenzene	18
m- and *p*-Ethyltoluene	19
α-Methylstyrene	20.5
Indane	24
Indene	25
m-Diethylbenzene	26
m-Ethylvinylbenzene	28
m-Divinylbenzene	30
Naphthalene	33
1-Methylnaphthalene	37.5
Biphenyl	40
Diphenylmethane	42
1,2-Diphenylethane	45

Note: Column packing — P1 = SE30 surface coating on extruded rolled metal capillary cylinder.
Sample — S1 = Dowex® 50W × 12, Hydrogen form.
Pyrolyzer type — CP = Curie Point iron capillary with quartz tube.

Blasius, E., Lohde, H., and Häusler, H., *Z. Anal. Chem.*, 264, 278, 1973.

Table P-GC 16
METHYL ACRYLATE — STYRENE COPOLYMER

Pyrolyzer type	F
Pyrolysis temperature (°C)	510
Pyrolysis time (sec)	n.a.
Sample size pyrolyzed (mg)	0.3—0.6
Temperature (°C)	100—260 @ 12°/min
Gas; flow rate (mℓ/min)	N_2; n.a.
Column	
Length, m	2
Diameter, mm I.D.	3
Material	Copper
Detector	FI

Compound	Relative Retention time (min)
Methanol	0.1
Methyl acrylate	1.1
Methyl methacrylate	1.5
Benzene	1.9
Toluene	2.0
Ethyl benzene	2.2
Styrene	3.0
α-Methylstyrene	3.8
β-Methylstyrene	4.1
CH_2-CH_2-CH_2 (with $COOCH_3$ on first and third C) $COOCH_3$ $COOCH_3$	7.1
MS hybrid dimers	10—12
SS dimers	13—16
MMM trimer	15.5
MMS hybrid trimers	17—19

Note: Compound abbreviation — M = Methyl acrylate.
S = Styrene.
Pyrolyzer type — F = Furnace.
Column packing — P1 = 5% FFAP on Diasolid L® (60/80 mesh).

Tsuge, S., Hiramitsu, S., Horibe, T., Yamaoka, M., and Takeuchi, T., *Macromolecules,* 8, 721, 1975.

Table P-GC 17
METHYL CELLULOSE AND ETHYL CELLULOSE

Sample	S1	S2	S3
Pyrolyzer type	C	R	C
Pyrolysis temperature (°C)	400	400	400
Pyrolysis time (sec)	10	10	10
Sample size pyrolyzed (mg)	1	1	1
Temperature (°C)		80°/1 min — 210° @ 20°/min	
Gas; flow rate (mℓ/min)			
Column			
Length, ft	6	6	6
Diameter, in. I.D.	1/8	1/8	1/8
Material	SS	SS	SS
Detector	FI	FI	FI
Compound		**Approximate retention time (min)**	
CO_2, CH_4	0.5	0.5	0.5
C_2H_6		1.0	
CH_3CHO		2.5	
CH_3OH	2.7		2.7
C_2H_5OH		3.5	
Acetone			5.7
$CH_3OC_2H_4OH$	6.2		6.2

Note: Sample — S1 = Methyl cellulose.
S2 = Ethyl cellulose.
S3 = Hydroxypropyl methyl cellulose.
Pyrolyzer type — C = Pyroprobe® CDS-120 Coil.
R = Pyroprobe® CDS-120 Ribbon.
Column packing — P1 = Chromosorb® 102 (100/120 mesh).

Chow, C. D., personal communication, 1978.

Table P-GC 18
METHYL METHACRYLATE — STYRENE COPOLYMER

Pyrolyzer type	F
Pyrolysis temperature (°C)	500
Pyrolysis time (sec)	n.a.
Sample size pyrolyzed (μg)	100—150
Temperature (°C)	50°—300° @ 10°/min
Gas; flow rate (mℓ/min)	He; 35
Column	
Length, m	2.5
Diameter, mm I.D.	3.5
Material	Glass
Detector	FI

Compound	Approximate retention time (min)
Methyl Methacrylate (M)	2
Styrene (S)	5
MS Dimer	15
SS Dimer	19
MSM Trimer	22
MSS Trimer	23.5
MSS Trimer	24
SMS Trimer	24.8
SSS Trimer	26.5

Note: Pyrolyzer type — F = Furnace (micro).
Column packing — P1 = 5% OV-17 on Diasolid H® (80/100 mesh).

Tsuge, S., Kobayashi, T., Nagaya, T., and Takeuchi, T., *J. Anal. Appl. Pyrol.*, 1, 133, 1979.

Table P-GC 19
NITROCELLULOSE

Pyrolyzer type	F
Pyrolysis temperature (°C)	500
Pyrolysis time (sec)	3—5
Sample size pyrolyzed (mg)	12—13
Temperature (°C)	110
Gas; flow rate (mℓ/min)	He; 120
Column	
Length, m	6
Diameter, in. O.D.	0.25
Material	n.a.
Detector	Thermistor

Compound	Approximate retention time (min)
Acetaldehyde	2.8
Methyl acetate	3.8
Methyl alcohol	4.5
Ethyl alcohol	5.2
Water	8.5

Note: Pyrolyzer type — F = Furnace.
Column packing — 10% Carbowax 1500® on Teflon 6® (<35 mesh).

Ettre, K. and Varadi, P. F., *Anal. Chem.*, 35, 69, 1963.

Table P-GC 20
6-66 NYLON HOMOPOLYMERS AND COPOLYMER

Sample	S1	S2	S3
Pyrolyzer type	F	F	F
Pyrolysis temperature (°C)	620	620	620
Pyrolysis time (sec)	n.a.	n.a.	n.a.
Sample size pyrolyzed (mg)	0.25	0.25	0.25
Temperature (°C)		50—200° @ 12°/min	
Gas; flow rate (mℓ/min)	N_2; 13	N_2; 13	N_2; 13
Column			
Length, m	2	2	2
Diameter, mm I.D.	3	3	3
Material	Cu	Cu	Cu
Detector	FI	FI	FI
Compound	**Approximate retention time (min)**		
Benzene	1.7	1.7	1.7
Acrylonitrile	2.0	2.0	2.0
Acetonitrile	2.1	2.1	2.1
Toluene	2.3	2.3	2.3
Pyrrolidine	3.0		3.0
Cyclopentanone		4.0	4.0
Pyridine	4.2	4.2	4.2
Styrene	4.9	4.9	4.9
Cyclohexanone	5.2	5.2	5.2
Pyrrole	7.2	7.2	7.2
ε-Caprolactam $(CH_2)_5CO$ / NH (ring)	14.0		14.0
Hexamethylenediamine		15.2	15.2

Note: Sample — S1 = 6-Nylon.
S2 = 66-Nylon.
S3 = 6-66 Nylon copolymer (56.8/43.2 mole %).
Column Packing — P1 = 5% Polyethylene glycol 6000 on Diasolid L® (80—100 mesh).

Senoo, H., Tsuge, S., and Takeuchi, T., *J. Chromatogr. Sci.*, 9, 315, 1971.

Table P-GC 21
PAINTS AND PLASTICS

Sample	S1	S2	S3
Pyrolyzer type	CP	CP	CP
Pyrolysis temperature (°C)	610	610	610
Pyrolysis time (sec)	10	10	10
Sample size pyrolyzed (μg)	14	7	44
Temperature (°C)	100—200° @ 8°/min		
Gas; flow rate (mℓ/min)	N_2; 60	N_2; 60	N_2; 60
Column			
Length, ft	5	5	5
Diameter, mm I.D.	4	4	4
Material	Glass	Glass	Glass
Detector	FI	FI	FI

Compound	Approximate retention time (min)		
Methane	0.4	0.4	0.4
Ethylene	0.8	0.8	0.8
Ethane	1.0	1.0	1.0
Methanol		2.2	
Propylene	2.5		2.5
Propane	2.9		2.9
Acetaldehyde	3.9	3.9	
Ethanol		5.2	
n-Butene	5.5		5.5
n-Butane	6.2		6.2
Acrolein	7.2		
Allyl alcohol	8.8		
Acetic acid			9.0
Methacrolein	10.4		
Benzene			13.2
Methyl methacrylate		15.2	
Toluene			18.5

Note: Samples — S1 = Linseed pentaerythritol-*o*-phthalate alkyd.
S2 = Acrylic emulsion.
S3 = Vinyl chloride-vinyl acetate copolymer (95—5).
Pyrolyzer type — CP = Curie Point.
Column packing — P1 = Porapak Q® (80—100 mesh).

May, R. W., Pearson, E. F., Porter, J., and Scothern, M. D., *Analyst*, 98, 364, 1973.

Table P-GC 22
PHENOLIC DERIVATIVES AND EPOXIDES

Sample	S1	S2	S3	S4
Pyrolyzer type	EO	EO	EO	EO
Pyrolysis temperature (°C)	550	550	550	550
Pyrolysis time (min)	10	10	10	10
Sample size pyrolyzed (mg)	2—3	2—3	2—3	2—3
Temperature (°C)		1.8°/min 90—130°		
Gas; flow rate (mℓ/min)	n.a.	n.a.	n.a.	n.a.
Column				
Length, m	1.4	1.4	1.4	1.4
Diameter, mm I.D.	3	3	3	3
Material	Glass	Glass	Glass	Glass
Detector	n.a.	n.a.	n.a.	n.a.

Compound	Relative Retention time (min)			
Phenol	7.9	7.9	7.9	7.9
2-Methylphenol	9.4	9.4	9.4	9.4
4-Methylphenol	11.3	11.3	11.3	11.3
2,4-Dimethylphenol	13.1	13.1	13.1	13.1
2-Ethylphenol		13.1	13.1	
4-Ethylphenol		16.2		16.2
4-Isopropylphenol		19.9		19.9

Note: Sample — S1 = Phenolic novolac.
S2 = Bisphenol A epoxide.
S3 = Epoxide novolac.
S4 = Polycarbonate.

Pyrolyzer type — EO = sealed glass ampoules heated in electric oven.

Column packing — HF-heated glass micro-beads (0.15—0.20 mm). impregnated with 0.01% H_3PO_4 and 0.13% 2,4-trixylenylphosphate.

Mlejnek, O. and Vorobjov, V., *J. Chromatogr.*, 79, 91, 1973.

Table P-GC 23
POLYACRYLONITRILE

Pyrolyzer type	F
Pyrolysis temperature (°C)	350
Pyrolysis time (sec)	30
Sample size pyrolyzed (mg)	0.001
Temperature (°C)	n.a.
Gas; flow rate (mℓ/min)	n.a.
Column	
Length, ft	4
Diameter, in. I.D.	0.2
Material	n.a.
Detector	FI

Compound	Approximate retention time (min)
Ammonia	2.0
HCN	3.0
Acetonitrile	4.2
Acrylonitrile	4.6
Methacrylonitrile	7.0

Note: Pyrolyzer type — F = Filament.
Column packing — P1 = 10% dinonylphthalate on Chromosorb W® (60/80 mesh).

Bell, F. A., Lehrle, R. S., and Robb, J. C., *Polymer,* 12, 579, 1971.

Table P-GC 24
POLYAMIDES

Sample	S1	S2	S3
Pyrolyzer type	CP	CP	CP
Pyrolysis temperature (°C)	610	610	610
Pyrolysis time (sec)	10	10	10
Sample size pyrolyzed (mg)	<1	<1	<1
Temperature (°C)	150	150	210
Gas; flow rate (mℓ/min)	N_2; 60	N_2; 60	N_2; 60
Column			
Length, m	2	2	2
Diameter, in. O.D.	0.25	0.25	0.25
Material	Cu	Cu	Cu
Detector	FI	FI	FI

Compound	Approximate retention time (min)		
Cyclopentanone	2.8		
Benzonitrile		7	
m-Phenylenediisocyanate			8
Caprolactam			9

Note: Sample — S1 = Polyamide 66® (Adipic acid, hexamethylene diamine).
S2 = Polyamide HT® (isophthallic acid, *m*-phenylenediamine).
S3 = Polyamide 6® (Caprolactam).
Pyrolyzer — CP = Curie Point.
Column packing — P1 = 10% Nitrilsilikon GE XE-60® on silanized Kieselgur®.

Kretzschmar, H.-J. and Gross, D., *Kunststuffe,* 65, 92, 1975.

Table P-GC 25
POLYBUTADIENE AND POLYBUTADIENE-STYRENE COPOLYMER

Sample	S1	S2
Pyrolyzer type	F	F
Pyrolysis temperature (°C)	600	600
Pyrolysis time (sec)	n.a.	n.a.
Sample size pyrolyzed (mg)	0.5—1	0.5—1
Temperature (°C)	25°/2 min — 200 @ 10°/min	
Gas; flow rate (mℓ/min)	N_2; 60	N_2; 60
Column		
Length, m	2	2
Diameter, in. I.D.	4	4
Material	Glass	Glass
Detector	FI	FI
Compound	**Retention time (min)**	
Methane	1.8	1.8
Ethane	3.2	3.2
Ethylene	4.2	4.2
Propane	6.6	6.6
Propylene	8.0	8.0
n-Butane	10.0	10.0
Butene-1,isobutene, *t*-butene-2	11.1	11.1
c-Butene-2	11.9	11.9
Butadiene[a]	3.0	13.0
C_5's	14—16.5	14—16.5
C_6's	17—20	17—20
Toluene	24	24
Vinylcyclohexene	26	26
Ethyl benzene		32
Styrene		39

Note: Sample — S1 = Polybutadiene.
S2 = Butadiene-styrene copolymer.
Pyrolyzer type — F = Furnace.
Column packing — P1 = 0.5% Carbowax 20M® on alumina (30/60 mesh) pretreated with 50% NaOH in methanol.

Host, M. and Deur-Šifter, D., *Chromatographia,* 5, 504, 1972.

Table P-GC 26
POLYCARBONATE

Column packing	P1
Pyrolyzer type	VT
Pyrolysis temperature (°C)	330
Pyrolysis time (sec)	n.a.
Sample size pyrolyzed (g)	1
Column temperature (°C)	70°—270° @ 24°/min
Gas; flow rate (mℓ/min)	n.a.
Column	
Length, m	n.a.
Diameter, mm I.D.	n.a.
Material	n.a.
Detector	FI

Compound	Approximate retention time (min)
Phenol	4.2
p-Methylphenol	5.2
p-Ethylphenol	6.0
p-Isopropylphenol	6.5
Diphenyl carbonate	10.2
4-Hydroxy-phenyl-4′phenyl-2,2 propane	11.5
Bisphenol A	14.5

Note: Column packing — P1 = 20% Apiezon K® on Kieselguhr.®
Pyrolyzer type — VT = Vacuum train; sample heated in molten salt bath.

Davis, A. and Golden, J. H., *J. Gas Chromatogr.*, 5, 81, 1967.

Table P-GC 27
POLYCARBONATE (MELT METHOD)

Pyrolyzer type	F
Pyrolysis temperature (°C)	580
Pyrolysis time (sec)	n.a.
Sample size pyrolyzed (mg)	0.1—0.5
Temperature (°C)	50—220° @ 20°/min
Gas; flow rate (mℓ/min)	N_2; 40
Column	
Length, m	2
Diameter, mm I.D.	3
Material	n.a.
Detector	FI

Compound	Retention time (min)
C_1-C_3 Hydrocarbons	1.0
Phenol	9.6
p-Cresol	10.0
p-Ethylphenol	10.8
Propylphenols (mostly isopropyl)	11.0
p-*t*-Butylphenol	11.7
Isopropenylphenol	12.9
	13.6

Note: Pyrolyzer type — F = Furnace.
Column Packing — P1 = 10% Carbowax 20M® on Diasolid L® (80/100 mesh).

Tsuge, S., Okumoto, T., Sugimura, Y., and Takeuchi, T., *J. Chromatogr. Sci.*, 7, 253, 1969.

Table P-GC 28
POLYCYCLOPENTADIENE

Packing	P1	P2
Pyrolyzer type	CP	CP
Pyrolysis temperature (°C)	770	770
Pyrolysis time (sec)	10	10
Sample size pyrolyzed (mg)	n.a.	n.a.
Temperature (°C)	65—160 @ 4°/min	180
Gas; flow rate (mℓ/min)	N_2; 23	N_2; 34
Column		
Length, ft	6	6
Diameter, in. I.D.	1/8	1/8
Material	SS	SS
Detector	FI	FI

Compound	**Retention time (min)**	
C_1	0.4	
C_2		0.6
C_3		0.8
C_4		1.3
Cyclopentadiene	1.1	2.8
Dicyclopentadiene		6.5
Benzene	1.6	
Toluene	2.3	
o-Xylene	3.4	
Styrene	4.6	
α-Methyl styrene	6.3	
Coumarone	7.0	
Indene	9.1	

Note: Column Packing — P1 = 5% XE60 on Gas Chrom Q® (80/100 mesh).
P2 = Porapak Q® (80/100 mesh).

Luke, B. G., *J. Chromatogr. Sci.*, 11, 435, 1973.

Table P-GC 29
POLYESTERS (AROMATIC)

Sample	S1	S2
Pyrolyzer type	CP	CP
Pyrolysis temperature (°C)	590	590
Pyrolysis time (sec)	5	5
Sample size pyrolyzed (mg)	0.1	0.1
Temperature (°C)	50°—250° @ 6°/min	
Gas; flow rate (mℓ/min)	N_2; 35	N_2; 35
Column		
Length, m	2	2
Diameter, mm I.D.	3	3
Material	Cu	Cu
Detector	FI	FI

Compound	Approximate retention time (min)	
Benzene	2	2
Vinylester of benzoic acid	18	
Allylcarbinolester of benzoic acid		22
Biphenyl	24	
Divinylester of terephthalic acid	32	
Diallylcarbiolester of terephthalic acid		37
Ethylene glycol dibenzoate	50	

Note: Sample — S1 = Polyethylene terephthalate.
S2 = Polybutylene terephthalate.
Pyrolyzer type — CP = Curie Point.
Column packing — P1 = 5% Thermon-1000® on Chromosorb WAW-DMCS® (80/100 mesh).

Sugimura, Y. and Tsuge, S., *J. Chromatogr. Sci.*, 17, 269, 1979.

Table P-GC 30
POLY(ETHYL ACRYLATE) POLYMER AND COPOLYMER

Sample	S1	S2
Pyrolyzer type	Coil	Coil
Pyrolysis temperaure (°C)	450	450
Pyrolysis time (sec)	n.a.	n.a.
Sample size pyrolyzed (mg)	0.025	0.025
Temperature (°C)	70	70
Gas; flow rate (mℓ/min)	N_2; 25	N_2; 25
Column		
Length, m	4	4
Diameter, mm I.D.	n.a.	na.
Material	n.a.	n.a.
Detector	n.a.	n.a.
Compound		
Ethanol	*	*
Methyl acrylate		*
Ethyl acrylate	*	*
Methyl methacrylate		*
Ethyl methacrylate		*

Note: Sample — S1 = Polyethyl acrylate.
S2 = Ethylacrylate-methacrylate copolymer (80/20).
Column packing — P1 = 20% silicone oil.

McCormick, H., *J. Chromatogr.*, 40, 1, 1969.

Table P-GC 31
POLYISOPRENE

Pyrolyzer type	CP
Pyrolysis temperature (°C)	500
Pyrolysis time (sec)	3
Sample size pyrolyzed (mg)	10
Temperature (°C)	115
Gas; flow rate (mℓ/min)	N_2; 27.5
Column	
Length, m	6
Diameter, mm O.D.	3.2
Material	n.a.
Detector	FI
Compound	**Approximate retention time (min)**
Isoprene	1.5
2,4-Dimethyl-4-vinylcyclohexene	15.5
1,4-Dimethyl-4-vinylcyclohexene	17
Diprene	25.5
Dipentene	28

Note: Pyrolyzer Type — CP = Curie Point.
Column Packing — P1 = Liquid crystal (4,4′-di-n-hexyloxyazoxybenzene) on Chromosorb W®.

Galin, M., *J. Chromatogr. Sci.-Chem.*, A7, 873, 1973.

Table P-GC 32
POLYISOPRENES

Sample	S1	S2
Pyrolyzer type	F	F
Pyrolysis temperature (°C)	350	350
Pyrolysis time (min)	30	30
Sample size pyrolyzed (g)	0.5—1.0	0.5—1.0
Temperature (°C)	100	100
Gas; flow rate (mℓ/min)	n.a.	n.a.
Column		
Length, m	n.a.	n.a.
Diameter, mm I.D.	n.a.	n.a.
Material	n.a.	n.a.
Detector	n.a.	n.a.

Compound	Approximate retention time (min)			
	P1	P2	P1	P2
1,3-Dimethylbenzene	7.5	7	7.5	7
1-Methyl-4-(1-methylethyl) cyclohexene	7.5	19	7.5	19
1-Methyl-4-(1-methylethenyl) cyclohexene	10	20		
1-Methyl-4-(1-methylethyl) benzene	14		14	19

Note: Sample — S1 = Natural rubber.
S2 = Synthetic *cis*-1,4-polyisoprene.
Pyrolyzer type — F = Furnace.
Column packing — P1 = 5% Poly(ethylene glycol) 4000 on Chromosorb B®
P2 = 10% SE30 on Chromosorb B®.

Gelling, I. R., Loadman, M. J., and Sidek, B. D., *J. Polymr. Sci.*, 17, 1383, 1979.

Table P-GC 33
POLY(ISOPROPENYL CYCLOHEXANE) AND COPOLYMERS

Sample	S1	S2	S3
Pyrolyzer type	C	C	C
Pyrolysis temperature (°C)	900	900	900
Pyrolysis time (sec)	20—30	20—30	20—30
Sample size pyrolyzed (μg)	50	50	50
Temperature (°C)		50°—250° @ 8°/min	
Gas; flow rate (mℓ/min)	n.a.	n.a.	n.a.
Column			
Length	n.a.	n.a.	n.a.
Diameter	n.a.	n.a.	n.a.
Material	n.a.	n.a.	n.a.
Detector	MS	MS	MS

Compound	Approximate retention time (min)		
Methylenecyclohexane + methyl-1-cyclohexene	1.7		1.7
Benzene		2.0	2.0
Ethylcyclohexane	2.2		
Toluene	2.7	2.7	2.7
Isopropenyl cyclohexane	3.2		
Ethylbenzene	3.8	3.8	
Cumene		4.4	4.4
Styrene		5.2	5.2
Isopropylcyclohexadiene	5.3		
α-Methylstyrene		6.8	6.8
Indene + phenylpropyne			8.4
Dicyclohexylpropane	9.9		
Methyl-1,2-dihydronaphthalene		11.3	
Methyleneindene			11.6
Naphthalene		12.3	12.3
2-Methylnaphthalene		13.9	13.9
Biphenyl		15.7	15.7
3-Cyclohexenyl-1-cyclohexyl-pentadiene	16.3		
Methylstilbene			18.7
Anthracene + phenanthrene		23.7	23.7

Note: Packing — P1 = 3% Dexsil 400® or 3% OV-101.
Sample — S1 = Poly(Isopropenylcyclohexane).
S2 = Poly(α-Methylstyrene).
S3 = Poly(Isopropenylcyclohexane/α-Methylstyrene); 33/67.
Pyrolyzer type — C = Coil probe with quartz tubes (CDS Pyroprobe 120)®.

Gritter, R. J., Gipstein, E., and Adams, G. E., *J. Polym. Sci.*, 17, 3959, 1979.

Table P-GC 34
POLY(METHACRYLIC ACID)

Pyrolyzer type	n.a.
Pyrolysis temperature (°C)	160—240
Pyrolysis time (sec)	3
Sample size pyrolyzed (mg)	5—20
Temperature (°C)	220
Gas; flow rate (mℓ/min)	He; 48
Column	
Length, m	2.5
Diameter, mm I.D.	3
Material	n.a.
Detector	Katharometer

Compound	Retention time (min)
Carbon dioxide	0.4
Water	0.6
Methacrylic acid	7.0

Note: Column packing — P1 = Porapak Q® (80—100).

Roux, F. X. and Audebert, R., *Analusis,* 2, 343, 1973.

Table P-GC 35
POLY(METHYL METHACRYLATE)

Pyrolyzer type	Pt. Coil
Pyrolysis temperature (°C)	1025
Pyrolysis time (sec)	5—10
Sample size pyrolyzed (mg)	2—3
Temperature (°C)	128
Gas; flow rate (mℓ/min)	He; 60
Column	
Length, ft	6
Diameter, in. O.D.	3/16
Material	n.a.
Detector	n.a.

Compound	Approximate retention
Gaseous products	1.4
Methanol	4.0
Ethanol	5.0
Methyl acrylate	7.3
Methyl propionate	9.0
Methyl isobutyrate	12.0
Methyl methacrylate	15.0

Note: Column packing — P1 = 30% Dinonylphthalate on firebrick (30—60 mesh).

Lehmann, F. A. and Brauer, G. M., *Anal. Chem.,* 33, 673, 1961.

Table P-GC 36
ISOTACTIC POLY(1-PENTENE)

Pyrolyzer type	MC
Pyrolysis temperature (°C)	115
Pyrolysis time (min)	75
Sample size pyrolyzed (mg)	n.a.
Temperature (°C)	n.a.
Gas; flow rate (mℓ/min)	n.a.
Column	
Length, m	n.a.
Diameter, mm I.D.	
Material	SS
Detector	TC

Compound	**Approximate retention time (min)**
CO_2	1
Acrolein	4
Acetaldehyde	7
Propionaldehyde	9
Acetone	10.5
Methanol	12.5
Ethanol	14
Butyraldehyde	16
Water	17
Propanol	17.5
Acetic acid	25
Propionic acid	27
Butyric acid	29

Note: Pyrolyzer type — MC = Mass Chromatograph (MC-2)®.
Column packing — P1 = 10% Carbowax 20M® on Chrom WAW® (80/100 mesh).

Gabbay, S. M. and Stivala, S. S., *Anal. Chem. Acta,* 78, 359, 1975.

Table P-GC 37
PARA — POLYPHENYLENE

Pyrolyzer type	Furnace
Pyrolysis temperature (°C)	615
Pyrolysis time (sec)	n.a.
Sample size pyrolyzed (mg)	n.a.
Temperature (°C)	150°/7 min — 250° at 15°/min
Gas; flow rate (mℓ/min)	n.a.
Column	
Length, m	1.52
Diameter, mm I.D.	6.4
Material	Glass
Detector	MS

Compound	Approximate retention time (min)
Biphenyl	3
Phenanthrene	20
m-Terphenyl	21
o-Terphenyl	28
p-Terphenyl	29
9,10-Benzphenanthrene	34
p-Quaterphenyl	46

Note: Column Packing — P1 = 3% OV-1 on Supelcoport® (80/100 mesh).

Fewell, L. L., *J. Chromatogr. Sci.*, 14, 564, 1976.

Table P-GC 38
POLYPROPYLENE

Pyrolyzer type	Furnace
Pyrolysis temperature (°C)	380
Pyrolysis time (sec)	n.a.
Sample size pyrolyzed (mg)	0.5—1
Temperature (°C)	25°/2 min — 200° @ 5°/min
Gas; flow rate (mℓ/min)	He; 60
Column	
Length, m	2
Diameter, mm O.D.	6
Material	n.a.
Detector	n.a.

Compound	Approximate retention time (min)
Methane	3
Ethane/Ethylene	5
Propane	9
Propylene	11
Isobutane	14
n-Butane	15
Butenes	16
n-Pentane	20
Pentenes	21
3-Methylpentane	24.5
Hexenes	26.5
2,4-Dimethylpentane	28
3-Methylhexane	29.5
Heptanes	30.5
4-Methylheptane	34
Isononane	41
Nonenes	44

Note: Column Packing — P1 = 5% Carbowax 20M® on Alumina® (30—60 mesh) treated with 50% NaOH in methanol.

Deur-Šifter, D. and Svob, V., *J. Chromatogr.*, 51, 59, 1970.

Table P-GC 39
POLYPROPYLENE (ISOTACTIC)

Pyrolyzer type	EHTB
Pyrolysis temperature (°C)	350
Pyrolysis time (sec)	n.a.
Sample size pyrolyzed (mg)	0.1
Temperature (°C)	n.a.
Gas; flow rate (mℓ/min)	n.a.
Column	
Length m	5
Diameter mm I.D.	2
Material	n.a.
Detector	FI

Compound	Approximate retention time (min)
Ethane	1.6
Propylene	1.8
Isobutylene	2.5
n-Pentane	4.3
2-Pentane	4.3
2-Methyl-1-pentene	7.0
2,4-Dimethyl-1-heptene	17.5

Note: Packing — P1 = 5% F-50(methyl-chlorophenyl silicone oil) on Chromosorb G.®
Pyrolyzer type — EHTB = Electrically heated Tantalum boat.

Blazsó, M. and Székely, T., *Eur. Polym. J.*, 10, 115, 1974.

Table P-GC 40
POLYSTYRENE

Pyrolyzer type	Pt. Coil
Pyrolysis temperature (°C)	1025
Pyrolysis time (sec)	5—10
Sample size pyrolyzed (mg)	2—3
Temperature (°C)	140
Gas; flow rate (mℓ/min)	He; 20
Column	
Length, ft	6
Diameter, in. O.D.	3/16
Material	n.a.
Detector	n.a.

Compound	Approximate retention time (min)
Ethylene/Acetylene	0.8
Benzene	2.2
Toluene	3.1
Ethylbenzene	4.8
Styrene	5.8

Note: Column Packing — P1 = 30% Apiezon L® on firebrick (30—60 mesh)

Lehmann, F. A. and Brauer, G. M., *Anal. Chem.*, 32, 673, 1961.

Table P-GC 41
POLYSTYRENE

Pyrolyzer type	CP
Pyrolysis temperature (°C)	760
Pyrolysis time (sec)	10
Sample size pyrolyzed (μg)	0.5
Temperature (°C)	n.a.
Gas; flow rate (mℓ/min)	n.a.
Column	
Length, m	20
Diameter, mm I.D.	0.26
Material	n.a.
Detector	FI

Compound	Approximate retention time (min)
Volatiles	1.5
Benzene	2.5
Toluene	3.8
Ethylbenzene	5.5
Styrene	7.5
Allylbenzene	8
α-Methylstyrene	9.2

Note: Pyrolyzer type — CP = Curie Point.
Column packing — P = SE52 glass open-tubular column.

Schmid, J. P., Schmid, P. P., and Simon, W., *Chromatographia,* 9, 597, 1976.

Table P-GC 42
HIGH IMPACT POLYSTYRENE

Pyrolyzer type	Furnace
Pyrolysis temperature (°C)	620
Pyrolysis time (sec)	20
Sample size pyrolyzed (mg)	1—2
Temperature (°C)	40°/2 min—170° @ 10°/min
Gas; flow rate (mℓ/min)	He; 60
Column	
Length, ft	4
Diameter, in I.D.	0.25
Material	SS
Detector	TC

Compound	Approximate retention
Light gases	1
1,3-Butadiene	2
Styrene	14

Note: Column packing — P1 = 20% SE30 on Chromosorb P® (60/80 mesh).

Armitage, F., *J. Chromatogr. Sci.,* 9, 245, 1971.

Table P-GC 43
IONICALLY CHLORINATED POLYSTYRENE

Sample	S1	S2	S3	S4	S5
Pyrolyzer type	F	F	F	F	F
Pyrolysis temperature (°C)	530	530	530	530	530
Pyrolysis time (sec)	60	60	60	60	60
Sample size pyrolyzed (mg)	0.5	0.5	0.5	0.5	0.5
Temperature (°C)			100—200° @ 5°/min		
Gas; flow rate (mℓ/min)	N_2	N_2	N_2	N_2	N_2
Column					
length, m	2	2	2	2	2
diameter, mm I.D.	3	3	3	3	3
material	n.a.	n.a.	n.a.	n.a.	n.a.
Detector	FI	FI	FI	FI	FI
Compound			**Approximate retention time (min)**		
Toluene	3				
Chlorobenzene					4.5
Styrene	5	5	5	5	5
p-Chlorotoluene					6.2
o-Chlorostyrene		10	10	10	10
p-Chlorostyrene		10.6	10.6	10.6	10.6
2,4-Dichlorostyrene			12	12	12
2,5-Dichlorostyrene			14.5	14.5	14.5
3,5-Dichlorostyrene			17	17	17

Note: Sample — S1 = Polystyrene.
S2 = Chlorinated polystyrene (Deg. Chlor. = 0.28).
S3 = Chlorinated polystyrene (Deg. Chlor. = 0.55).
S4 = Chlorinated polystyrene (Deg. Chlor. = 0.98).
S5 = Chlorinated polystyrene (Deg. Chlor. = 1.61).
Pyrolyzer type — F = Furnace.
Column packing — P1 = 10% Carbowax 6000® on Diasolid L® (80/100 mesh).

Tadaoki, O. and Tsugio, T., *Makromol. Chem.*, 167, 305, 1973.

Table P-GC 44
POLY(STYRENE-BUTADIENE-METHYLMETHACRYLATE)

Packing	P1	P1
Sample	S1	S2
Pyrolyzer type	CP	CP
Pyrolysis temperature (°C)	600	600
Pyrolysis time (sec)	10	10
Sample size pyrolyzed (mg)	0.5—1.5	0.5—1.5
Temperature (°C)	110°/6 min—210° @ 6°/min	
Gas; flow rate (mℓ/min)	He; 30	He; 30
Column		
Length, m	3	3
Diameter, mm I.D.	3.1	3.1
Material	SS	SS
Detector	FI	FI

Compound	Approximate retention time (min)	
Butadiene dimer (BB) + methylmethacrylate (M)	1.0	1.0
Toluene	1.3	1.3
Ethylbenzene	1.5	1.5
Styrene	2.8	2.8
Methyl styrene	4.0	4.0
Butadiene trimer (BBB)	4.8	4.8
Dimer BM	6.5	6.5
Trimer MBB		12
Styrene-butadiene hybrid dimer (SB)		13
Styrene-butadiene hybrid dimer (SB)	13.5	
Trimer BBM	14	14
Methyl methylmethacrylate		15
Trimer SBB	16.5	
Methylmethacrylate-styrene dimer (MS)	17	
Trimer SBB	18	
Trimer SBB	19	
Diphenylpropane	21	21
Styrene dimer (SS)	24	24

Note: Packing — P1 = 10% Carbowax 20M® on Chromosorb W® (60/80 mesh).
Sample — S1 = Butadiene-styrene (40:10) backbone grafted with styrene (25) and methylmethacrylate (25).
S2 = Butadiene-methylmethacrylate (40:10) backbone grafted with styrene (35) and methyl methacrylate (15).
Pyrolyzer type — CP = Curie Point.

Eustache, H., Robin, N., Danial, J. C., and Carrega, M., *Eur. Polym. J.,* 14, 239, 1977.

Table P-GC 45
AROMATIC POLYSULPHONES

Packing	P1	P2
Pyrolyzer type	F	F
Pyrolysis temperature (°C)	450	450
Pyrolysis time (sec)	n.a.	n.a.
Sample size pyrolyzed (mg)	n.a.	n.a.
Temperature (°C)	115	115
Gas; flow rate (mℓ/min)	He; 50	H; 50
Column		
Length, m	1.5	2
Diameter, mm I.D.	4	4
Material	n.a.	n.a.
Detector	TC	TC

Compound	**Approximate retention time (min)**	
Hydrogen	1	
CO	1.3	
CH_4	1.9	
CO_2	3	
CO + CH_4		1.1
CO_2		1.2
SO_2		2.1

Note: Pyrolyzer type — F = Furnace connected in vacuum loop.
Column packing — P1 = Porapak Q®.
P2 = SKT Charcoal®.

Laktionov, V. M. and Zhurauleva, I. V., *Polym. Sci., USSR,* 17, 3232, 1975.

Table P-GC 46
POLY(VINYL ACETATE) AND VINYL ACETATE/1-DODECENE COPOLYMER

Sample	S1	S2
Pyrolyzer type	PF	PF
Pyrolysis temperature (°C)	650	650
Pyrolysis time (sec)	n.a.	n.a.
Sample size pyrolyzed (mg)	0.5	0.5
Temperature (°C)	100	100
Gas; flow rate (mℓ/min)	He; 30	He; 30
Column		
Length, m	100	100
Diameter, mm I.D.	0.25	0.25
Material	SS Cap.	SS Cap.
Detector	MS	MS

Compound	**Approximate retention time (min)**	
Propane/Carbon dioxide	14	14
Butadiene/Butene	15	15
Acetone	15.5	15.5
3-Methyl-1,3-butadiene		16
1-ene-3-Pentyne	16.5	16.5
1-Hexene		17.5
Acetic acid	18	18
Benzene	19	19
5-Methyl-1-hexene		21
1-Octene		28
Ethylbenzene	37	37
m-Xylene	40	40
p-Xylene	40.5	40.5
Styrene	43	43
o-Xylene	44	44
1-Nonene		45
n-Propylbenzene	58	58
m-Methyl-styrene	70	70
p-Methyl-styrene	71	71
o-Methyl-styrene	74	74
1-Decene		75
β-Methyl-styrene	87	87

Note: Sample — S1 = Poly(vinyl acetate).
S2 = Vinyl acetate/1-dodecene copolymer.
Pyrolyzer type — PF = U-shaped platinum filament.
Column packing — P1 = Squalene coating on walls.

Sellier, N., Jones, C. E. R., and Guiochon, G., *J. Chromatogr. Sci.,* 13, 383, 1975.

Table P-GC 47
POLYVINYL TRIMETHYL SILANE (ISOTACTIC)

Pyrolyzer type	EHTB
Pyrolysis temperature (°C)	350
Pyrolysis time (sec)	n.a.
Sample size pyrolyzed (mg)	0.1
Temperature (°C)	n.a.
Gas; flow rate (mℓ/min)	n.a.
Column	
Length, m	3.5
Diameter, mm I.D.	2
Material	n.a.
Detector	FI

Compound	**Approximate retention time (min)**
Tetramethyl silane	1.8
Trimethyl vinyl silane	2.2
2-Trimethyl silyl propylene	3.2
1,3-Trimethyl silyl propane	9
1,3-Trimethyl silyl propylene	9
2,4-Trimethyl silyl butene-1	10.5
2,4,6-trimethyl silyl hexene-1	16.5

Note: Packing — P1 = 5% OV-1 on Chromosorb W®.
Pyrolyzer — EHTB = Electrically heated Tantalum boat.

Blazsó, M. and Székely, T., *Eur. Polym. J.*, 10, 115, 1974.

Table P-GC 48
RUBBER IN COMPOUNDED CURED STOCK

Sample	S1
Pyrolyzer type	Furnace
Pyrolysis temperature (°C)	700
Pyrolysis time (sec)	30
Sample size pyrolyzed (μg)	500—800
Temperature (°C)	35
Gas; flow rate (mℓ/min)	He; 30
Column	
Length, ft	40
Diameter, in. I.D.	3/16
Material	Aluminum
Detector	FI

Compound	Approximate retention time (min)
Methane	8
Ethane + ethene	8.5
Propane	10
Propene	11.2
2-Methylpropane	12
Propadiene + butane	13.8
1-Butene + 2-methylpropene	15
trans-2-Butene	18
cis-2-Butene	20
1,3-Butadiene	21
3-Methyl-1-butene	23
1-Pentene	29.5
2-Methyl-1-butene	31.5
trans-2-Pentene	35
cis-2-Pentene	36
2-Methyl-2-butene	40
Isoprene	46
Highers	Backflushed

Note: Sample — S1 = Mixture 40% natural rubber, 30% styrene-butadiene rubber, 30% ethylene-propylene-terpolymer.

Column packing — 10% Tricresylphosphate on Chromosorb P® (60/80 mesh).

Krishen, A., *Anal. Chem.*, 44, 494, 1972.

Table P-GC 49
STYRENE — ISOPRENE COPOLYMER

Pyrolyzer type	CP
Pyrolysis temperature (°C)	601
Pyrolysis time (sec)	10
Sample size pyrolyzed (mg)	n.a.
Temperature (°C)	85
Gas; flow rate (mℓ/min)	He; 12
Column	
Length, ft	50
Diameter, in. I.D.	0.02
Material	n.a.
Detector	FI

Compound	Approximate retention time (min)
Isoprene	2.2
Cyclohexadiene + benzene	3.1
Methylcyclohexadiene	4.1
Toluene	5.1
Cyclooctadiene	6.0
Isopropylcylohexene	6.5
Ethylbenzene	8.6
p-Xylene	9.4
m-Xylene	10.5
o-Xylene	12.0
Isopropylbenzene	13.0
Styrene	15.0
1,4-Dimethyl-1-vinylcyclohexene	16.2
Diprene + C_3 benzene	24.6
Dipentene	28.2
Dimethylcyclooctadienes + isopropenyl benzene	34.0

Note: Pyrolyzer type — CP = Curie Point.
Column packing — P1 = SCOT containing 5-ring polyphenylether liquid phase with Apiezon L® and Igepal CO880®.

Jackson, M. T. and Walker, J. Q., *Anal. Chem.*, 43, 74, 1971.

Table P-GC 50
SYNTHETIC RUBBERS

Sample	S1	S2	S3	S4
Pyrolyzer type	F	F	F	F
Pyrolysis temperature (°C)	650	650	650	650
Pyrolysis time (sec)	5—10	5—10	5—10	5—10
Sample size pyrolyzed (μg)	100	100	100	100
Temperature (°c)	45°/5 min — 160° @ 6°/min			
Gas; flow rate (mℓ/min)	Ar; 20	Ar; 20	Ar; 20	Ar; 20
Column				
Length, m	3	3	3	3
Diameter, mm I.D.	3	3	3	3
Material	SS	SS	SS	SS
Detector	FI	FI	FI	FI
Compound	**Approximate retention time (min)**			
Butadiene	2	2	2	2
Isoprene	3			
Vinylcyclohexene		11	11	11
Dipentene	15			
Styrene			17	
Methylstyrene				20

Note: Sample — S1 = Polyisoprene.
S2 = Polybutadiene.
S3 = Styrene-butadiene rubber.
S4 = Methylstyrene-butadiene rubber.
Pyrolyzer type — F = Filament.
Column packing — P1 = 15% Reoplex 400® on Chezasorb AW® (0.25—0.316 mm).

Alekseeva, K. V., *J. Anal. Appl. Pyrol.*, 2, 19, 1980.

Table P-GC 51
VINYL CHLORIDE — VINYL ACETATE COPOLYMERS

Sample	S1	S2	S3
Pyrolyzer type	F	F	F
Pyrolysis temperature (°C)	430	430	430
Pyrolysis time (sec)	60	60	60
Sample size pyrolyzed (mg)	0.5—2	0.5—2	0.5—2
Temperature (°C)		120° — 230° @ 7°/min	
Gas; flow rate (mℓ/min)	N_2; 55	N_2; 55	N_2; 55
Column			
Length, m	0.5	0.5	0.5
Diameter, mm I.D.	3	3	3
Material	SS	SS	SS
Detector	FI	FI	FI
Compound		**Approximate retention time (min)**	
Acetic acid		7	7
Benzene	12	12	12
Toluene	16	16	

Note: Sample — S1 = Polyvinyl chloride (PVC).
S2 = VC-Ac Copolymer® (5.5% VAc).
S3 = Polyvinyl acetate (PVAc).
Pyrolyzer type — F = Furnace.
Column packing — P1 = Porapak Q® (120/150 mesh).

Okumoto, T., Takeuchi, T., and Tsuge, S., *Bulletin of the Chemical Society of Japan,* 43, 2080, 1970.

Table P-GC 52
VINYL ACETATE — OLEFIN COPOLYMERS

Sample	S1	S2
Pyrolyzer type	EHPS	EHPS
Pyrolysis temperature (°C)	650	650
Pyrolysis time (sec)	na	na
Sample size pyrolyzed (mg)	0.5	0.5
Temperature (°C)	100	100
Gas; flow rate (mℓ/min)	He; Split	He; Split
Column		
Length, m	100	100
Diameter, mm I.D.	0.25	0.25
Material	SS	SS
Detector	MS	MS

Compound	Approximate retention time (min)	
Propane + carbon dioxide	13	13
Butadiene + butene	14	14
Acetone	15	15
3-Methyl-1,3-butadiene		16
1-Hexene		17
Acetic acid	18	18
Benzene	20	20
5-Methyl-1-hexene		20.5
Toluene	27	27
1-Octene		29
Ethylbenzene	38	38
m-Xylene	40	40
p-Xylene	40.5	40.5
Styrene	42	42
o-Xylene	44	44
1-Nonene		45
n-Propylbenzene	58	58
m-Methylstyrene	69	69
p-Methylstyrene	70	70
o-Methylstyrene	74	74
1-Decene		76

Note: Packing — P1 = Capillary column; wall coated with squalane.
Sample — S1 = Poly(vinyl acetate).
S2 = Vinyl acetate-dodecene-1 copolymer.
Pyrolyzer — EHPS = Electrically heated platinum sheet (u-shape).

Sellier, N., Jones, C. E. R., and Guiochon, G., *J. Chromatogr. Sci.*, 13, 383, 1975.

Table P-GC 53
VINYLIDENE CHLORIDE — VINYL CHLORIDE HOMOPOLYMERS AND COPOLYMERS

Sample	S1	S2	S3
Pyrolyzer type	F	F	F
Pyrolysis temperature (°C)	430	430	430
Pyrolysis time (sec)	n.a.	n.a.	n.a.
Sample size pyrolyzed (mg)	0.1—0.3	0.1—0.3	0.1—0.3
Temperature (°C)		50—210°C @ 16°/min	
Gas; flow rate (mℓ/min)	N_2; n.a.	N_2; n.a.	N_2; n.a.
Column			
Length, m	2	2	2
Diameter, mm I.D.	3	3	3
Material	n.a.	n.a.	n.a.
Detector	FI	FI	FI
Compound		**Approximate retention time (min)**	
Vinylidene chloride		1.8	1.8
Benzene	3.5		3.5
Chlorobenzene			6.5
1,3-Dichlorobenzene			8.7
1,3,5-Trichlorobenzene		9.5	9.5

Note: Sample — S1 = Polyvinyl chloride.
S2 = Polyvinylidene chloride.
S3 = Vinylidene chloride — vinylchloride copolymer (28 mole % vinylidene chloride).
Pyrolyzer type — F = Heated Furnace.
Column packing — P1 = 10% Carbowax 6000® on Diasolid L® (80—100 mesh).

Tsuge, S., Okumoto, T., and Takeuchi, T., *Makromol. Chem.*, 123, 123, 1969.

Table P-GC 54
ARYLSULFONIC ACIDS

Sample	S1	S2
Pyrolyzer type	TTM	TTM
Pyrolysis temperature (°C)	710—790	710—790
Pyrolysis time (sec)	n.a.	n.a.
Sample size pyrolyzed (mg)	1.5	1.5
Temperature (°C)	70—220° @ 26°/min	
Gas; flow rate (mℓ/min)	n.a.	n.a.
Column		
Length, ft	7	7
Diameter, in. O.D.	1/8	1/8
Material	SS	SS
Detector	TC	TC

Compound	Approximate retention time (min)	
Water	2.1	2.1
Sulfur dioxide	2.9	2.9
Benzene	6.0	6.0
Toluene	7.7	

Note: Sample — S1 = *p*-Toluenesulfonic acid.
S2 = Benzenesulfonic acid.
Pyrolyzer type — TTM = Tube-type microreactor.
Column packing — P1 = Porapak Q® (80—100 mesh).

Siggia, S. and Whitlock, L. R., *Anal. Chem.*, 42, 1719, 1970.

Table P-GC 55
Di-t-BUTYL PEROXIDE

Pyrolyzer type	IP
Pyrolysis temperature (°C)	310
Pyrolysis time (sec)	n.a.
Sample size pyrolyzed (mg)	n.a.
Temperature (°C)	65
Gas; flow rate (mℓ/min)	He; 54
Column	
Length, ft	6
Diameter, in. O.D.	0.25
Material	SS
Detector	TC

Compound	Approximate retention time (min)
Ethane	0.7
Acetone	2.3
Benzene	4.0

Note: Pyrolyzer type — IP = Injection port of gas chromatograph.
Column packing — P1 = 20% Carbowax 1500® on Chromosorb P®.

Hyden, S., *Anal. Chem.*, 35, 113, 1963.

Table P-GC 56
PIGMENT DYE

Sample	S1
Pyrolyzer type	CP
Pyrolysis temperature (°C)	760
Pyrolysis time (sec)	10
Sample size pyrolyzed (μg)	1
Temperature (°C)	6°C/min 25—250°
Gas; flow rate (mℓ/min)	n.a.
Column	
Length, m	20
Diameter, mm I.D.	0.28
Material	n.a.
Detector	FI

Compound	Approximate retention time (min)
Benzene	1.2
Toluene	2
Chlorobenzene	3
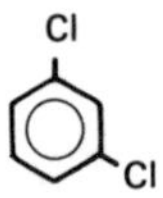	6
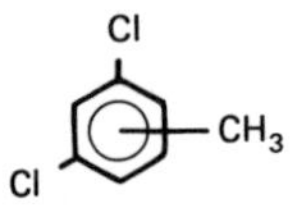	8.8
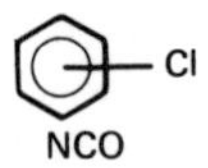	9.2
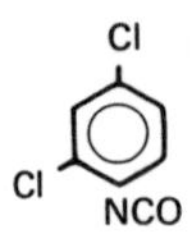	12.8
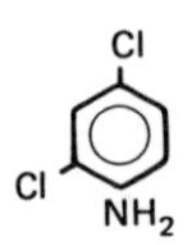	14.2
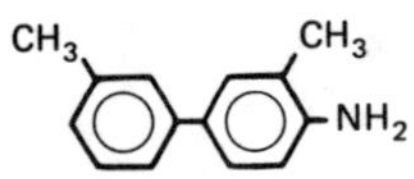	25

Note: Sample — S1 =

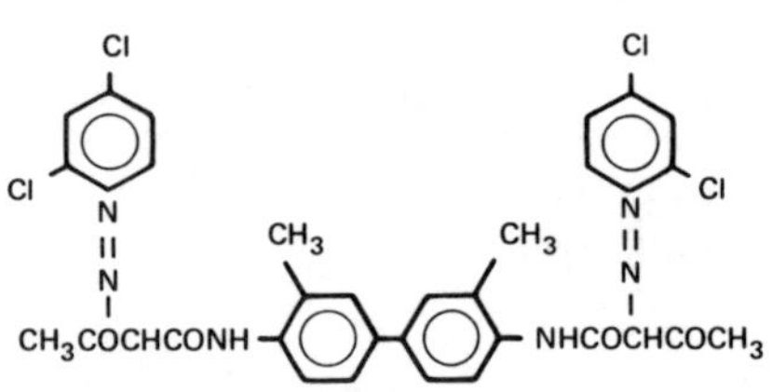

Pyrolyzer type — CP = Curie Point.
Column packing — P = SE52 glass open — tubular column.

Schmid, J. P., Schmid, P. P., and Simon, W., *Chromatographia,* 9, 597, 1976.

IV.3 LIQUID COLUMN CHROMATOGRAPHY TABLES

INTRODUCTION

For the most part, tables are arranged in alphabetical order by compound or polymer class. The reader is referred to the List of Tables in Section III or the master index for specific compounds.

Please note n.a. = not available from the original source; R_t = retention time; V_e = elution volume in mℓ and solvent composition always v/v unless otherwise stated.

Table LC-1
ACRYLAMIDE AND ACRYLIC ACID IN POLYACRYLAMIDE

Column	30 × 1.2 cm DOWEX 50W-X8, 200—400 mesh, H^+ form
Eluent	Water
Flow rate	2.5 mℓ/min
Amount	0.5 mℓ containing 5 mg polymer
Detector	Ultraviolet at 200 nm

Compound	Rt (min)
Polyacrylamide	6
Acrylic acid	8
Acrylamide	16

Schmotzer, G., *Chromatographia,* 4, 391, 1971.

Table LC-2
ACRYLAMIDE MONOMER IN POLYACRYLAMIDE AND ENVIRONMENTAL SAMPLES

Column	25 × 0.46 cm Whatman Partisil-10 ODS-2®
Eluent	Distilled water
Flow rate	2 mℓ/min
Injection volume	20 μℓ
Detector	Ultraviolet at 208 nm, 0.02 AUFS

For aqueous solution and aqueous impinger samples — filter and direct injection.

For polyacrylamide — 5 g of polyacrylamide was extracted with 50 mℓ of 80-20 methanol-water for 3 to 4 hr on a mechanical shaker. 20 μℓ of the clear phase was injected.

Compound	Rt (min)
Acrylic acid	1.4
β-hydroxypropanamide	2.1
Acetamide	3.0
Acrylamide	5.4
Propanamide	7.3
Acrylonitrile	11.8
Methacrylamide	18.0
Butanamide	20.8
Methacrylonitrile	46.0

Skelly, N. E. and Husser, E. R., *Anal. Chem.,* 50, 1959, 1978.

Table LC-3
ACRYLAMIDE MONOMER IN NATURAL AND POLLUTED AQUEOUS ENVIRONMENTS

Column	25 × 0.4 cm Spherisorb 10μ ODS
Eluent	distilled water
Flow rate	4 mℓ/min
Injection volume	12 μℓ
Detector	Ultraviolet at 196 nm

α,β-Dibromopropionamide retention time: 4.0 min

A 100-mℓ aliquot of sample was extracted with 20 mℓ of ethyl acetate and 20 mℓ of hexane to remove organics. 0.5 mℓ of HBr and 15 g of KBr were added to the aqueous phase with stirring. Saturated bromine water (5.0 mℓ) was added with stirring and the flask set aside overnight in the dark at 0°C to allow formation of the α,β-dibromopropionamide drivative. The excess bromine was decomposed by adding 1 *M* sodium thiosulfate solution dropwise. Anhydrous sodium sulfate (30 g) was added while using a magnetic stirrer to effect vigorous stirring.

The brominated sample was extracted with three portions of ethyl acetate (15 mℓ) and dried with 2 g sodium sulfate. The ethyl acetate was evaporated to 3 mℓ and transferred to a glass vial and evaporated to dryness. The residue was taken up with 120 μℓ of water and a 12 μℓ aliquot was chromatographed. Sensitivity was 0.2 μg/1.

Brown, L. and Rhead, M., *Analyst,* 104, 391, 1979.

Table LC-4
ACRYLAMIDE MONOMER IN AQUEOUS DISPERSED-PHASE POLYMER SYSTEMS

Procedure: The acrylamide is extracted from the aqueous dispersion by the following "extraction-coagulation" procedure. Ten mℓ of the methylene chloride, 40 mℓ of water and 0.5 mℓ of concentrated hydrochloric acid are placed into the 4 oz bottle and stirred vigorously on a magnetic stirrer. Five g of the dispersed polymer is added dropwise and allowed to continue stirring for 30 min. The samples are then centrifuged at 10,000 rpm and 0°C for 2 hr in a refrigerated centrifuge. In order to achieve the detection limit of 0.1 ppm, decant off the supernate (40 mℓ) and concentrate to 10 mℓ by evaporation with filtered air. If the concentrated extracts are not clear, filter through a 10μ filter.

The extracts are chromatographed by ion exclusion liquid chromatography under the following conditions:

Column	Dowex® resin 50W-X4 (250 × 9 mm)
Eluent	0.01N H_2So_4
Injection	0.5 mℓ
Flow rate	0.7 mℓ/min
Pressure	100 psi
Detector	225 nm
Acrylamide	27 min

Husser, E. R., Stehl, R. H., Price, D. R., and DeLap, R. A., *Anal. Chem.,* 49, 154, 1977.

Table LC-5
ACRYLAMIDE MONOMER IN NON-AQUEOUS DISPERSED-PHASE POLYMER SYSTEMS

Column	Partisil 10 PAC (250 × 4.6 mm)
Eluent	15% methanol, 85% methylene chloride
Flow rate	1 mℓ/min
Pressure	500 psi
Detector	240 nm
Acrylamide	3.7 min
Benzamide	4.2 min

Procedure: The residual acrylamide monomer is extracted by adding 10g (±0.1 g) of the nonaqueous dispersed polymer dropwise to 50 mℓ of methanol containing the internal standard benzamide. The mixture is stirred for 2hr on a magnetic stirrer. It is then centrifuged for 15 to 20 min to isolate the polymer from solution. A 10-μℓ aliquot of the supernate is injected directly onto the column and separated according to the chromatographic conditions.

Husser, E. R., Stehl, R. H., Price, D. R., and DeLap, R. A., *Anal. Chem.,* 49, 154, 1977.

Table LC-6
ACRYLAMIDE SYNTHESIS PROCESS ANALYSIS

Column	30 × 0.4 cm Water's Associates μBondapak C_{18}®
Eluent	30:70 methanol:water or water
Flow rate	1 mℓ/min
Injection volume	10 μℓ
Detector	R401 differential refractometer

Compound	**Retention time (min)**	
	water	**30:70 MeOH/water**
β-hydroxypropionitrile	4.1	3.5
Acrylamide	4.9	
Acrylic acid	7.3	
Acrylonitrile	8.4	4.4

Onuoha, N. I., Chaplain, R. P., and Wainwright, M. S., *Chromatographia,* 12, 709, 1979.

Table LC-7
ACRYLIC ACID MONOMER IN NATURAL AND POLLUTED AQUEOUS ENVIRONMENTS AND POLYACRYLATES

Column	25 × 0.4 cm Whatman PXS 10/25 PAC®
Eluent	0.01% v/v orthophosphoric acid in D.I. water
Flow rate	4 mℓ/min
Injection volume	12 μℓ
Detector	Ultraviolet at 195 nm, 0.01 AUFS
Acrylic acid retention time	5.5 min

For aqueous samples - filter and direct injection. For polyacrylates and acrylic acid copolymers - a 0.5 g sample was added to 50 mℓ of 1:1 methanol-water and allowed to stand overnight. An aliquot of this extract was injected into the liquid chromatographic column.

Brown, L., *Analyst,* 104, 1165, 1979.

Table LC-8
ANTIOXIDANTS

Column	Zorbax® — SIL 4.6 mm × 25 cm (DuPont)®
Eluent	Hexane/methanol (99.8:0.2)
Flow rate	1 mℓ/min
Detection	Ultraviolet detector

Compound	Rt
BHT, 2,6-di-t-butyl-*p*-cresol	1.1
Ionox 220, 4,4′-methylenebis(2,6-di-t-butylphenol	1.4
Ionox 330	2.2
Irganox 1076 (octadecyl-3-(3,5-di-t-butyl-4-hydroxy-phenyl) propionate	2.8
AO 425, 2,2′-methylenebis (4-ethyl-6-t-butylphenol	3.3
A02246	3.5

Wims, A. M. and Swarin, S. J., *J. Appl. Polym. Sci.*, 19, 1256, 1975.

Table LC-9
GPC SEPARATION OF ANTIOXIDANTS

Column set C	4 Styragel columns Porosity 250, 100, 60, and 60A
Eluent	THF
Flow rate	1 mℓ/min
Detection	Refractive index

Trade Name	Chemical name	MW	V_e[a]
Irganox 1010®	tetrakis[methylene-3-(3′,5′-di-t-butyl-4′-hydroxyphenyl)propionate] methane	1168	18.88
Ionox 330®	1,3,5-trimethyl-2,4,6-tris(3,5-di-t-butyl-4-hydroxybenzyl)benzene	775	19.50
Topanol CA®	1,1,3-tris(5-t-butyl-4-hydroxy-2-methylphenyl)butane	554	20.65
Santowhite® powder	4,4′-butylidenebis(3-methyl-6-t-butylphenol	382	21.75
A02246	2,2′-methylenebis(4-methyl-6-t-butylphenol	340	23.00
Santovar®	2,5-di-t-amylhydroquinone	250	23.75
Tenamene®	2,6-di-t-butyl-4-methylphenol	220	24.87
Agerlite alba®	hydroquinone monobenzyl ether 4-butylcatechol	200	25.19
Ansul HA®	hydroquinone monomethyl ether	124	27.06
DSTDP	distearyl 3,3′-thiodipropionate	682	19.83
DLTDP	dilauryl 3,3′-thiodipropionate	514	20.69

V_e[a] is the peak elution volume in counts. One count = 5 mℓ.

Wims, A. M. and Swarin, S. J., *J. Appl. Polym. Sci.*, 19, 1243, 1975.

Table LC-10
EPOXY RESINS

Column	Bondapack C_{18} Corasil,® 2 mm × 4 ft Waters Associates®
Eluent	Gradient, 100% water to 100% THF
Flow rate	1 mℓ/min
Amount	0.8 mg
Detection	Ultraviolet
Separated	EPI-REZ 510 Celanese
	EPON, 836, 828, 1001, 1004 Shell Chem.®
	DER 331, 511, 661 Dow Chem.®
	ARALDITE 6010 Ciba-Geigy®
	ERL 2774 Union Carbide®

Dark, W. A., Conrad, E. C., and Crossman, L. W., Jr., *J. Chromatogr.*, 91, 260, 1974.

Table LC-11
PHENOL-FORMALDEHYDE RESINS

Column	Merkosorb SI-60,® 10μ 0.2 mm × 0.5 m
Eluent	dichloromethane/dioxane/methanol (500:2:4) alternate eluents dichloromethane/dioxane (100:3) dichloromethane/dioxane/methanol (100:1:1)
Flow rate	0.75 mℓ/min
Detection	Ultraviolet and refractive index

Compound	Relative absorption factor @ 254 nm	Relative retention
Phenol	1.00	1.00
Bis-(2-hydroxy-phenol)methane	0.95	1.07
2-hydroxybenzyl alcohol	0.83	2.10
Bis-(4-hydroxyphenol)methane	0.96	2.50
4-Hydroxybenzyl alcohol	0.83	6.30
3-Hydroxybenzyl alcohol	0.84	7.60

Sebenik, A. and Lapanje, S., *J. Chromatogr.*, 106, 454, 1975.

Table LC-12
PHTHALATE PLASTICIZERS

Column	Corasil I,® 0.5% β,β′ oxydipropionitrile on 37-50μ support, 2.1 mm I.D. 1000 mm
Eluent	Isooctane
Flow rate	0.5 mℓ/min
Detection	Ultraviolet

Compound	Rt (min)
Didecyl phthalate	6.3
Decylbenzyl phthalate	10.2
Dibenzyl phthalate	23.5

Majors, R. J., *J. Chromatogr. Sci.*, 8, 343, 1970.

Table LC-13
POLYESTER PREPOLYMER OLIGOMERS

Sample preparation:
15 mg sample in 3-dram vial
Add 5 mℓ chloroform/ethanol (9:1) and stir magnetically
Inject 10μℓ of solution

Column	Zorbax-SIL® 2.1 mm I.D. × 25 cm (DuPont)®
Eluent	Chloroform/ethanol 1250:12
Pressure	2000 psi constant pressure
Detection	Ultraviolet at 254 nm

Poly(ethylene terephthalate) oligomer	Rt (min)
Heptamer	12.5
Hexamer	14.1
Petamer	15.5
Tetramer	18.0
Trimer	21.4
Dimer	28.6
Monomer	42

Zaborsky, L. M., II., *Anal. Chem.*, 49, 1166, 1977.

Table LC-14
POLYETHYLENE GLYCOL IN NON-IONIC SURFACE ACTIVE AGENTS

Column	1.6 cm I.D. × 50 cm water jacketed at 35°C
Packing	Mix in small quantities 30g Whatman Grade CF-11® cellulose powder with 10 mℓ 30% aqueous NaCl solution and add to a ethyl acetate-filled column
Eluent	Elute with 510 mℓ ethyl acetate and collect 150 mℓ 60 mℓ and 300 mℓ fractions. The fractions are evaporated to dryness and the weights recorded
Sample	Add 2 g of sample to the column and wash in with 5 mℓ of acetate

The reaction product is contained in the first fraction and the polyethylene glycol in the third. This should be determined through appropriate experimentation for various materials.

Taylor, E. A., *J. Chromatogr.*, 64, 71, 1972.

Table LC-15
POLYOXYETHYLATED PHENOLS

Column	Porasil A® (60) 37-75μm (Waters Associates)® 1/8 in. × 2 ft (anal.) 3/8 in. × 8 ft (prep.)
Eluent	Linear gradient, A-ethyl acetate-acetic acid-water (10:32:30) B-ethyl acetate
Flow rate	0.5 mℓ/min (anal.) 6 mℓ/min (prep.)
Detection	Ultraviolet 280 nm

Sample of Triton X-100® (*p-tert*-octylphenoxy-polyethoxyethanol oligomers) was separated into 22 components.

Allen, C. F. and Rice, L. I., *J. Chromatogr.*, 110, 151, 1975.

Table LC-16
RUBBER ADDITIVES

Column	Kopolymer ST-DVB No. VIII polystyrene gel® (United Chemical and Metalurgical Works, Production Company, Utsi, Czechoslovakia) grain size 0.040—0.056 mesh
Eluent	THF
Flow rate	30—50 mℓ/min
Pressure	2—4 atm
Temperature	25°C
Amount	Samples 2—5% in THF
Detection	Ultraviolet and refractive index

Trade Name	Structure	Mol. wt.	V_e (cm³)	V_r
Antioxidant 116®	Phenyl-2-naphthylamine	219.27	190.6	0.653
PBN®	Phenyl-2-naphthylamine	219.27	190.6	0.653
UOP 288®	*N*,N′-Bis-(1-methylheptyl)-*p*-phenylenediamine	332.58	149.1[a]	0.834
Santoflex IP®	*N*-Phenyl-N′-isopropyl-*p*- phenylenediamine	226.34	172.8[a]	0.720
Nonox ZA®	*N*-Phenyl-N′-isopropyl-*p*- phenylenediamine	226.34	172.8[a]	0.720
UOP 688®	*N*-Phenyl-N′-octyl-*p*- phenylenediamine	296.47	156.0[a]	0.797
Altofane DIP®	*N*,N′-Diphenyl-*p*-phenylenediamine	260.36	169.8[a]	0.733
JZF®	*N*,N′-Diphenyl-*p*-phenylenediamine	260.36	169.8[a]	0.733
Antioxidant 123®	*N*,N′-Di-2-naphthyl-*p*-phenylenediamine	360.46	161.0	0.733
DNPD®	*N*,N′-Di-2-naphthyl-*p*-phenylenediamine	360.46	161.0	0.733
Thermoflex A®	2:1:1 mixture of phenyl-2-naphthylamine,			
	4,4′-dimethoxydiphenyl-amine and		167.9	0.741
	N,N′-diphenyl-*p*-phenylenediamine		187.6[b]	0.663
Age Rite Hipar®	2:1:1 mixture of phenyl-2-naphthylamine,		167.9	0.741
	p-isopropoxydiphenylamine and		176.8	0.704
	N,N′-diphenyl-*p*-phenylene-diamine		187.6[a]	0.663

[a] Contains some other compounds.
[b] The presence of 4,4′-dimethoxydiphenylamine was not proved.

Protivová, J., Pospisil, J., and Holcik, J., *J. Chromatogr.*, 92, 361, 1974.

Table LC-17
STABILIZERS IN TETRAETHYLENE GLYCOL DIMETHYLACRYLATE

Column	Si-10 Micropak,® 1.5 mm I.D. × 25 cm (Varian)®
Eluent	Isooctane/ethyl acetate/methylene chloride (7.6:2.1:2)
Flow rate	1 mℓ/min
Amount	1 μℓ neat
Detection	Ultraviolet 292 nm

Stabilizer	Rt (min)
4,4′-thiobis (3-methyl-6-*tert*-butylphenol)	1.7
Methyl ether of hydroquinone	2.8
Hydroquinone	9.6

Pasteur, G. A., *Anal. Chem.,* 49, 363, 1977.

Table LC-18
FREE TOLUENE DIISOCYANATE IN POLYURETHANE PREPOLYMERS

Column	50 × 0.2 cm Corasil II® in stainless steel
Eluent	Isooctane/dioxane (4:1 v/v)
Flow rate	1 or 3 mℓ/min
Amount	4 μℓ 1—10% prepolymer in dioxane; standard 0.4% in dioxane
Detection	Ultraviolet and refractive index

Compound	Time (min)
TDI	1.8
Diol-TDI adduct	2.3
Triol-TDI adduct	3.7

McFayden, P. E., *J. Chromatogr.,* 123, 468, 1976.

Table LC-19
GEL CHROMATOGRAPHIC BEHAVIOUR OF VULCANIZATION ACCELERATORS AND OTHER SELECTED RUBBER ADDITIVES

Trade name	Structure	Mol. wt.	V_e	V_r
Hermat ZDM®	Zinc *N*-dimethyldithio-carbamate	305.82	203.4[a]	0.612
Hermat ZDK®	Zinc *N*-diethyldithio-carbamate	361.94	181.7[a]	0.685
Hermat FEDK®	Zinc *N*-ethylphenyl-dithiocarbamate	458.02	166.9[a]	0.745
	Nickel *N*-dimethyldithio-carbamate	299.16	211.3[c]	0.589
	Nickel *N*-dibutyldithio-	467.48	156.0[a]	0.797
	Sodium *N*-diethyldithio-	319.56	184.7[a,c]	0.673
Hermat Cu®	Copper *N*-dimethyldithio-	303.99	207.4[a,c]	0.600
Rodanin S-62®	2-Mercaptoimidazoline	102.16	203.4[a]	0.612
Pneumax MBT®	2-Mercaptobenzothiazole	167.24	200.5[a]	0.620
Antioxidant MB®	2-Mercaptobenzimidazole	150.20	185.6[a]	0.670
	2-Phenylbenzothiazole	211.29	218.2	0.570
Pneumax DM®	Dibenzothiazyl disulphide	332.8	209.3[a,c]	0.594
Hermat TMT®	Tetramethylthiuram disulphide	240.45	219.2	0.567
Pentazine BAFD®	Di-*o*-benzamidiphenyl-disulphide	458.62	169.0	0.736
Hermat Zn MBT®	Zinc salt of 2-mercapto-benzothiazole	231.61	200.5[a,c]	0.620
Sulfenax CB/30®	Benzothiazyl-2-cyclo-hexylsulphenamide	264.41	188.6[a]	0.659
Santocure Moor®	Benzothiazyl-2-morpholyl-sulphenamide	267.37	208.4	0.597
Herax UTS®	Activated hexamethylene-tetramine	140.19	183.7[a,c]	0.677
ISAF-black®				
Triumf®	Mineral oil			
	Zinc oxide			
	Sulphur		304.1[a]	0.409
Paraffin®				
Stearin®			152.1	0.817
			157.0	0.792

Note: V_e = elution volume (cm^3).
V_r = relative elution volume related to the exclusion limit ($V_r = V_o/V_e$).
V_o = 124.5 cm^3.

[a] Partly insoluble in acetone and THF.
[b] Only UV detection.
[c] Contains some other compounds. Insoluble in acetone and THF. Only diffuse UV detection in the range 120—220 cm^3.

Protivová, J., Pospisil, J., and Holcik, J., *J. Chromatogr.*, 92, 361, 1974.

IV. 4 THIN-LAYER CHROMATOGRAPHY

INTRODUCTION

For the most part, tables are arranged in alphabetical order by compound classification. The reader is referred to the list of tables in Section III or the master index for specific compounds.

Please note n.a. = not available from the original source; technique is ascending development, unless otherwise stated; solvent composition always v/v, unless otherwise stated; if solid is dissolved in solvent, g/v.

Table TLC-1
ACCELERATORS IN RUBBER COMPOUNDS

Layer	Polygram Sil-G UV_{254} (Machery Nagel®)
Solvent	Benzene/ethyl acetate/acetone (100:7:2)
Detection	D1 D2 D1 + D2

D1 = Iodoplatinate solution; 3 mℓ 10% platinum chloride plus 97 mℓ water to which is added 100 mℓ 6% aqueous KI solution

D2 = Dibromobenzoquinone chlorimide; 1% solution dibromobenzoquinone chlorimide in methanol, 4% solution sodium hypochlorite in water, and 1% sodium bicarbonate in water

Accelerators

Chemical	Trade Name or Abbreviation	D1	D2	D1 + D2	R_f
Diphenylguanidine	DPG®	Brown, white edge	Blue	Green, white edge	0.00
Di-*o*-tolyguanidine	DOTG®	Brown, white edge	Blue	Green, white edge	0.00
2-Mercaptobenzothiazole	MBT®	Yellow	Orange	Rust, white edge	0.30
Benzothiazyl disulfide	MBTS®	Cream streak	Pale yellow	Orange streak, white edge	0.50
Tetramethylthiuram disulfide	TMTD®	Yellow, brown, white edge	Dark red, pink edge	Red, white edge	0.45
Tetramethylthiuram monosulfide	TMTM®	Yellow, brown, white edge	Dark red, pink edge	Yellow, red edge	0.45
N,N-Dicyclohexylbenzothiazole-2-sulfenamide	Vulcacit DZ®	Yellow, white edge	Yellow orange	Orange, white edge	0.70
Phosphorus-containing poly-sulfide	Vulcadone 3 SN®	White, gray edge	Yellow	Rust, white edge	0.85
N-Cyclohexyl-2-benzothiazole sulfenamide	Santocure®	Yellow, pale edge	Yellow orange	Orange, white edge	0.63

Table TLC-1 (continued)
ACCELERATORS IN RUBBER COMPOUNDS

Accelerators

Chemical	Trade Name or Abbreviation	D1	D2	D1 + D2	R_f
N-tert-Butyl-2-benzothiazole sulfenamide	Santocure NS®	Yellow, pale edge	Yellow orange	Orange, white edge	0.55
Zinc diethyl-dithiocarbamate	ZDC®	Yellow	Brown, red edge	Rust, white edge	0.73
Zinc salt of 2 mercaptobenzo-thiazole	ZMBT®	Yellow, white	Red	Rust	0.25
2-Benzothiazyl-*N,N*-diethylthiocarbamyl sulfide	Ethylae®	Yellow, white edge	Yellow orange	Rust, white edge	0.38
4-Morpholine disulfide	Morfax®	White	Yellow	Brown, white edge	0.35
Copper dimethyl-dithiocarbamate	Cumate®	Pale brown streak	Red streak	Brown streak	0.20—0.70
Dipentamethylene thiuram tetra-sulfide	Tetrone A®	Brown streak, cream center	Orange yellow	Red streak	0.30—0.80
2-(Morpholino-thio)-benzothiazole	Santocure MOR®	Yellow, cream	Yellow	Orange, white edge	0.30, 0.63
Hexamethylene tetramine	Vulcacit H 30®	Red, white edge	Yellow	Red, white edge	0.00
Morpholine disulfide	Sulfasan R®	Grey, white edge	Yellow	Rust, white edge	0.38,0.20
2-Mercapto imidazoline	Na 22®	White	Brown	Brown orange	0.00
2,2′-Dibenzamido-diphenyl disulfide	Pepton 22®	Cream, brown edge	Pale orange	Red, white edge	0.35

Millinger, M. B., ***Anal. Chem.*****, 46, 746, 1974.**

Table TLC-2
AMINE ANTIOXIDANTS IN HIGH-TEMPERATURE JET ENGINE LUBRICANTS

Layer	(E. Merck®) silica gel 60 F-254
Solvents	heptane/toluene (1:1)
Detection	densitometer (after 16 hr for color development)
Amount	1 $\mu\ell$ (sample diluted 3 to 1 with heptane)

Compound	R_F
Phenyl-alpha-naphthylamine	0.59
p,p′-Dioctyldiphenylamine	0.75
N,p-octylphenyl-alpha-naphthylamine	0.84

Sniegoski, P. J., ***J. Chromatogr. Sci.*****, 15, 328, 1977.**

Table TLC-3
AMINE RESIDUES FROM SULFINAMIDE ACCELERATORS IN RUBBER VULCANIZATES

1g of vulcanizate (10-20 mg pieces), 1 mℓ of 10% $NaHCO_3$, and 1 mℓ of 1% 4-chloro-7-nitrobenzo-2, 1,3-oxadiazole (NBD-Cl) in methyl ethyl ketone are heated to 80° for 30 min. To the cooled mixture add 2 mℓ toluene and extract. Evaporate the toluene layer under vacuum or in a stream of nitrogen. Dissolve the residue in 0.2 mℓ chloroform for TLC examination. For standards use appropriate amines (10 mg).

Layer	Kieselgel GC_{254} (E. Merck)® 0.4 mm
Solvent	Petroleum ether (b.p. 40°—60°C) twice followed by toluene/ethyl acetate (4:1 v/v)
Detection	UV and visible light

Rubber vulcanizates	R_F
Morpholine	0.18
2,6-Dimethylmorpholine *cis*	0.25
2,6-Dimethylmorpholine *trans*	0.35
tert-Butylamine	0.62
Cyclohexylamine	0.68

Higgins, G. M. C. and McSweeney, G. P., *Rubber Chemistry and Technology,* 47, 1206, 1974.

Table TLC-4
ANTIOXIDANT ANALYSIS

Layer	1:1 Kieselgel G (Merck®) and aluminum oxide, Type E (Merck®), 0.3 mm
Solvent	chloroform
Amount	5 μℓ mg/mℓ
Detection	spray with 5% iodine in methanol

	Observed R_f and R_t values of antioxidant			
	As a single substance		As one of a mixture	
Antioxidant	R_f	R_t	R_f	R_t
1,1,3-*tris*(2-methyl-4-hydroxy-5-*t*-butylphenyl)	0.25	0.25	0.25	0.25
4,4′-thiobis (6-*t*-butyl-*m*-cresol)	0.43	0.44	0.51	0.52
N,N-*di*-*β*-naphthyl-*p*-phenylenediamine	0.87	0.89	0.90	0.92
2,2′-methylenebis[6-(1-methylcyclohexyl)]-*p*-cresol	0.95	0.97	0.95	0.97
2,6-*di*-*t*-butyl-*p*-cresol	0.98	1.00	0.98	1.00
2,2′-thiobis(4-methyl-6-*t*-butylphenol)				

Miles, D. T., *Analyst,* 99, 724, 1924.

Table TLC-5
ANTIOXIDANTS IN PLASTIC FORMULATIONS

Layer	Kiesilgel GF_{254}; 0.25 mm (E. Merck®)		
Detection	D1		
Amount	2 μℓ of 1% solution in chloroform		
Solvent	S1	S2	S3
Compound		R_F	
Nonox SP®	SF	0.61,0.93	0.50
Nonox TBS®	SF	0.98	0.62
Nonox WSP®	SF	0.85	0.52
Nonox EX®	SF	0.56	0.38
Nonox WSL®	SF	0.82	0.54
BHT®	SF	0.98	0.91
2246®	SF	0.75	0.56
Nonox Cl®	SF	0.65,0.84	0.49
Nonox DPPD®	SF	0.66	0.48
Nonox OD®	SF	0.56	0.40
Nonox ZA®	SF	0.43,0.38	0.38
Santoflex 75®	SF	0.25,0.61,0.74	0.47,0.54
Santoflex AW®	SF	0.80,0.51,0.55	0.54,0.45
Santoflex R®	SF	0.26,0.53	0.39
DLTDP®	SF	0.61	0.41
Nonox NS®	SF	0.52	0.4
Superlite®	SF	0.64,0.93	0.58
Polygard®	SF	0.52,0.61	0.53
Nonox HO	SF	0.46,0.64,0.60, 0.99	0.52
Nonox WSO	SF	0.58	0.39
Irganox 1076	SF	0.57,0.75	0.56

Sample Preparation — Weigh 2 samples of plastic material (2—5 g) into Soxhlet extractor and extract 8 hr with ether. Evaporate the ether and add 5 mℓ ethanol; filter quantitatively if insoluble polymer remains. Evaporate ethanol to dryness then dissolve residue in 5 mℓ chloroform.

Note: Solvent — S1 = Butanol-glacial acetic acid (97:3).
S2 = Benzene-ethyl acetate-acetone (100:5:2).
S3 = Chloroform-hexane (2:1).

Detection — D1 = View plate under UV light, then spray with 0.2% 2,6-dichloro-*p*-benzoquinone-4-chloroimine in absolute ethanol. Heat for a few minutes then expose layer to iodine vapor for 5 min.
SF = Solvent front

Sampson, D. and Currel, B. R., *Analyst,* 96, 515, 1971.

Table TLC-6
ANTIOXIDANTS IN RUBBER COMPOUNDS

Layer	Polygram Sil-G UV_{254} (Machery Nagel®)
Solvent	Benzene:ethyl acetate:acetone (100:7:2)
Detection	D1, D2, D1 + D2 D1 = Iodoplatinate solution; 3 mℓ 10% platinum chloride plus 97 mℓ water to which is added 100 mℓ 6% aqueous KI solution. D2 = Dibromobenzoquinone chlorimide; 1% solution dibromobenzoquinone chlorimide in methanol, 4% solution sodium hypochlorite in water, and 1% sodium bicarbonate in water.

Antioxidants

Chemical	Trade name or abbreviation	D1	D2	D1 + D2	R_f
Acetone diphenylamine condensation product	Nonox BL®		Pale green	Green blue	0.53
Butraldehyde aniline condensation product	Antox®		Brown yellow edge	Grey purple	0.40
4,4′-Butylidene-*bis*-(6-*tert*-butyl-3-methylphenol)	Santowhite® powder		Purple	Grey	0.37
4-Phenyl-*N*-cyclohexyl-*p*-phenylene diamine	Antioxidant 400®	Grey blue	Purple	Purple	
2,2′-Methylene-*bis*-4-ethyl-6-*tert*-butyl-phenol	Antioxidant 425®		Green white ctre	Rust yellow ctre	0.58
Diphenylamine acetone reaction product	Ble®	Green blue (weak)	Yellow	Rust	0.48
N-Phenyl-*N*′-1,3-dimethyl butyl *p*-phenylenediamine	Santoflex 13®	Blue brown	Brown yellow edge	Purple	0.48
6-Dodecyl-1,2-dihydro-2,2,4-trimethyl-quinoline	Santoflex DD®		Yellow	Rust yellow ctre	0.38
Diphenyl-*p*-phenylene-diamine	DPPD®	Green	Purple	Purple	0.47
Mixture of diaryl *p*-phenylenediamines	Wingstay 100®	Brown green	Brown yellow edge	Purple	0.50
6-Ethoxy-1,2-dihydro-2,2,4-trimethyl-quinoline	Santoflex AW®		Cyclamen	Purple	0.40
N,N-*Bis*(1-ethyl-3-methylpentyl)-p-phenylenediamine	UOP88®	Cyclamen	Cyclamen	Cyclamen	
N,N-*Bis*(1-methyl-heptyl)-*p*-phenylene-diamine	Santoflex 217®	Purple	Pink, grey edge	Purple	0.25
N,N-*Bis*(1,4 dimethyl pentyl)-*p*-phenylene-diamine	Santoflex 77®	Purple	Brown	Purple	0.23
2,2′-Methylene-*bis*-(6 ∇-methylcyclohexyl-4-methylphenol)	Nonox WSP®		Yellow	Yellow orange edge	
4-Isopropylaminodi-phenylamine	Nonox ZA®	Grey green	Brown yellow edge	Purple	0.33
A substituted phenol	Nonox WSL®		Rust	Yellow purple edge	0.55
β-Naphthylamine			Brown	Purple	0.22
Tri(nonylated phenyl)-phosphite	Polygard®	White	Purple, orange	White purple edge	0.57, 0.33
Substituted styrenated phenol	Wingstay S®		Purple, blue	Purple	0.60, 0.37
2,6-di-*tert*-Butyl-4-methylphenol	Ionol®		Rust	Yellow brown edge	0.90
2,2′-Methylene-*bis*(4-methyl 6-*tert*-butyl-phenol)	Antox 2246®		Yellow	Orange brown edge	0.57

Table TLC-6 (continued)
ANTIOXIDANTS IN RUBBER COMPOUNDS

	Antioxidants				
Chemical	**Trade name or abbreviation**	**D1**	**D2**	**D1 + D2**	**R_f**
Modified phenyl β-naphthylamine	Nonox DN®		Pink	Purple	0.53
A phenol condensation product	Nonox EX®		Yellow	Yellow Purple streak	0.38
A blend of acrylamines	Nonox HFN®		Pink	Purple	0.53
Sym. di-β-naphthyl-*p*-phenylenediamine	Nonox CI®		Brown streak	Lavender streak	0.00
Phenyl-β-naphthylamine + diphenyl-*p*-phenylenediamine	Nonox HP®	Weak pale green	Yellow	Purple	0.47, 0.53
A phenol condensation product	Nonow WSP®		Yellow	Rust	0.37
Octylated diphenyl-amines	Octamine®		Pale pink	Brown green	0.99
2,2,4-Trimethyl-1,2-dihydroquinoline	Flectol Flakes®	(Weak response)	Turquoise	Green	0.45, 0.35, 0.20, 0.13, 0.05
A phenol condensation product	AFD®		Grey		0.60
N-N'-*Bis*(1-methyl heptyl)-*p*-phenylene-diamine	UOP288®	Cyclamen	Pink,blue	Purple	0.50, 0.08

Note: Color reactions given in tables refer to pure compounds. At the dilution level in a rubber stock, many of the positive responses given by antioxidants to the iodoplatinate reagent are weak and, in many instances, could be regarded as negative. Those antioxidants which remain clearly evident at such dilution are Santoflex® 13, Santoflex® 77, Santoflex® 217, and Nonox ZA.®

Millinger, M. B., *Anal. Chem.,* 46(6), 746, 1974.

Table TLC-7
ETHYLENE OXIDE ADDUCTS OF 4-*TERT*-OCTYLPHENOL

Layer	Silica gel G (E. Merck®)
Amount	10—20 μg
Detection	Dragondorff's reagent
Solvent	Ethyl acetate:acetic acid:water (140:32:30)

Ethoxymer no.	**R_f**
7	0.69
8	0.59
9	0.50
10	0.41
11	0.32
12	0.28
13	0.27
14	0.21
15	0.16

Allen, C. F. and Rice, L. I., *J. Chromatogr.,* 110, 151, 1975.

Table TLC-8
2-HYDROXYETHYL METHACRYLATE IMPURITIES

Layer Silica gel 60 F_{254} (E. Merck®)
Detection UV; Iodine vapor, and/or potassium permanganate

Solvent	S1	S2	S3	S4	S5
Compound			R_F		
Ethylene glycol	0.02	0.03	0.02	0.04	0.02
2-Hydroxyethyl methacrylate	0.23	0.27	0.21	0.33	0.33
Methacrylic acid	0.36	0.73	0.64	0.55	0.55
Ethylene dimethacrylate	0.65	0.79	0.71	0.79	0.79

Note: Solvent — S1 = Hexane-diethyl ether (1:1).
S2 = Hexane-diethyl ether (1:1, saturated with 25% HNO_3)
S3 = Sulfuric acid impregnated silica gel; Hexane-diethyl ether (1:1)
S4 = Hexane-methyl isobutyl ketone-*n*-octanol (9:2:1)
S5 = Hexane-methyl isobutyl ketone-*n*-octanol (9:2:1, saturated with 25% HNO_3)

Brinkman, U. A. Th., Van Schalk, T. A. M., DeVries, G., and deVisser, A. C., *J. Chromatogr.*, 115, 195, 1975.

Table TLC-9
SHORT CHAIN α-METHYLSTYRENE POLYMERS

Layer Kieselgel $HF_{254+366}$
Detection Fluorescence
Solvent Petroleum ether (b.p. 40°—60°C)

Compound	R_F
α-Methylstyrene dimers	0.30
α-Methylstyrene trimers	0.15
α-Methylstyrene tetramers	0.10

Carver, F. W. S., Gallacher, E. J., and Tunstall, F. I. H., Chromatographia, 3, 333, 1970.

Table TLC-10
PHENOLIC PYROLYSIS PRODUCTS

Layer	Silica gel (Silufol®, Kavalier®)
Detection	Diazonium salt of sulphanilic acid (0.1 g in 20 mℓ of 10% sodium carbonate solution.
Solvent	Chloroform

Compound	R_f
2,6-Dimethylphenol	0.68
Phenol	0.20
2-Methylphenol	0.34
4-Methylphenol	0.20
3-Methylphenol	0.21
2,4-Dimethylphenol	0.34
2-Ethylphenol	0.40
2,5-Dimethylphenol	0.37
2,3-Dimethylphenol	0.35
4-Ethylphenol	0.21
2-Isopropylphenol	0.44
3,5-Dimethylphenol	0.23
3,4-Dimethylphenol	0.23
3-Isopropylphenol	0.25
4-Isopropylphenol	0.22
2,3,5-Trimethylphenol	0.38
4-*tert*-Butylphenol	0.24

Mlejnek, O. and Vorobjov, V., *J. Chromatogr.*, 79, 91, 1973.

Table TLC-11
ULTRAVIOLET ABSORBERS IN PLASTIC FORMULATIONS

Layer	Kiesilgel GF_{24}; 0.25 mm (E. Merck®)		
Detection	D1		
Amount	2 μℓ of 1% solution in chloroform		
Solvent	S1	S2	S3
Compound		**R_F**	
Eastman DOBP	SF	0.76	0.39
Uvinul 400	SF	0.22	0.08
Permyl B100	SF	0.24	0.08
Salol	SF	0.73	0.45
Tinuvin HE	SF	SF	0.71
Tinuvin P	SF	0.76	0.59
Eastman OPS	SF	0.79	0.65
UV 318	SF	0.26	0.14
Cyasorb 1988	SF	0.26	0.12
Cyasorb 1084	SF	0.24	0.08
Uvinul N35	SF	0.28	0.16

Sample Preparation — Weigh two samples of plastic material (2—5 g) into Soxhlet® extractor and extract 8 hr with ether. Evaporate the ether and add 5 mℓ ethanol; filter quantitatively if insoluble polymer remains. Evaporate ethanol to dryness then dissolve residue in 5 mℓ chloroform.

Note: Solvent — S1 = Butanol-glacial acetic acid (97:3).
S2 = Benzene-ethyl acetate-acetone (100:5:2).
S3 = Chloroform-hexane (2:1).

Detection — D1 = View the plate under UV light, then spray with 2% borax solution in 50% ethanol-water followed by 2,6-dichloro-p-benzoquinone-4-chlorimine solution.
SF = Solvent front.
NL = Not located with reagent listed.

Sampson, D. and Currel, B. R., *Analyst,* 96, 515, 1971.

Table TLC-12
ORGANOTIN STABILIZERS IN PLASTIC FORMULATIONS

Layer	Kiesilgel GF_{254}; 0.25 mm (E. Merck®)			
Detection	D1			
Amount	2 $\mu\ell$ of 1% solution in chloroform			
Solvent		S1	S2	S3

Compound	R_F		
Dibutyltin dilaurate	0.29	NL	NL
Dioctyltin dilaurate	0.46	NL	NL
Butyltin trichloride	0.0	0.0	0.0
Dimethyltin dichloride	0.04	NL	NL
Diphenyltin dichloride	0.21,0.76	NL	NL
Hexbutylditin	0.26,0.96	SF	SF
Tributyltin laurate	0.82	NL	NL
Dibutyltin-*bis*-(2-ethylhexyl-thioglycollate)	0.33	NL	NL

Sample Preparation — Weigh two samples of plastic material (2—5 g) into Soxhlet extractor and extract 8 hr with ether. Evaporate the ether and add 5 mℓ ethanol to the flask. Filter quantitatively to remove undissolved polymer. Evaporate ethanol to a very small volume, then dilute to 5 mℓ with ethanol.

Note: Solvent — S1 = Butanol-glacial acetic acid (97:3).
S2 = Benzene-ethylacetate-acetone (100:5:2).
S3 = Chloroform-hexane (2:1).
Detection — D1 = View the plate under UV light, then spray with 2% borax solution in 50% ethanol-water followed by 2,6-dichloro-*p*-benzoquinone-4-chlorimine solution.
SF = Solvent front.
NL = Not located with reagent listed.

Sampson, D. and Currel, B. R., *Analyst*, 96, 515, 1971.

Table TLC-13
POLYETHYLENE GLYCOLS

Layer		L1	L2
Detection	sprayed with iodine solution		
Amount	10—20 μg		
Solvent		S1	S2

Polyethylene Glycol	$\overline{Mw}/\overline{Mn}$[a]	R_F	
1500	< 1.1	0.61	0.46
2000	< 1.1	0.51	0.44
3000	< 1.1	0.40	0.39
6000	1.08	0.19	0.36
10000	< 1.1	0.03	0.26
28000	< 1.1	0.00	0.20

Note: Layer — L1 = Silica gel F_{254} (Brinkmann Instruments)®.
L2 = Aluminum oxide F_{254}.
Solvent S1 = Ethylene glycol-methanol (4:1).
S2 = Gradient-methanol to methanol/dimethyl formamide (4:1).

[a] Based on GPC Data.

Otocka, E. P. and Hellman, M. Y., *Macromolecules*, 3(3), 362, 1970.

Table TLC-14
MOLECULAR WEIGHT DISTRIBUTION OF POLYETHYLENE GLYCOLS

Layer	Silica gel G 20—40 μ (E. Merck®) 0.2 mm
Detection	Iodine vapor followed by densitometry
Amount	10 μℓ 2% w/w in chloroform
Solvent	S1 S2

Derivatization — the polyethylene glycol sample was dried for 1 hr at 60° C and 10 mm Hg, 0.5g of sample was dissolved in 0.5 mℓ of warm anhydrous pyridine in a 10-mℓ test tube. The solution was cooled to 30° C and 0.2 mℓ of thionyl chloride was added dropwise over a period of 30 min, with continuous agitation. An additional, 0.4 mℓ portion of thionyl chloride was added quickly and the mixture was heated for three hours at 80 ± 2°C. The mixture was then extracted with at least eight 10-mℓ portions of ethyl ether, then taken up with 5—10 mℓ of water and the resulting aqueous solution was extracted with two 5-mℓ portions of ether. The ether extracts were combined and distilled first at atmospheric pressure and then at 10 mm Hg. A 2% (w/v) solution of this was prepared and 10 μℓ was applied to the TLC plate.

	R_F VALUE	
Ethoxymer no.	S1	S2
5	0.72	
6	0.66	
7	0.61	0.84
8	0.55	0.80
9	0.49	0.74
10	0.42	0.68
11	0.36	0.60
12	0.30	0.52
13	0.25	0.44
14	0.21	0.36
15	0.16	0.30
16	0.11	0.24
17		0.19
18		0.14
19		0.11
20		0.08
21		0.06

Note: S1 = 2-Butanone/water (90:10).
S2 = 2-Butanone/water/chloroform/methanol (87.9:9.7:1.4:1.9).

Fauretto, L., Marletta, G. P., and Gabrielle, L. F., *J. Chromatogr.*, 50, 304, 1970.

Table TLC-15
POLYETHYLENE AND POLYPROPYLENE GLYCOL

Layer	Neutral silica gel
Detection	0.05% (w/w) ethanolic solution of rhodamine B
Amount	5 μℓ 1% solutions in industrial methylated spirits
Solvent	S1 S2

Derivatization — 200 mg of polyglycol 20 mℓ benzene containing 500 mg of 3,5-dinitrobenzoyl chloride, 0.2 mℓ of pyridine and two or three antibumping granules were placed in a 100-mℓ round-bottomed flask. The contents of the flask were refluxed for 1 hr, cooled and transferred to a 100 mℓ separatory funnel. The flask was rinsed with three 5-mℓ portions of benzene, which were added to the separatory funnel; 25 mℓ of 0.1N sulphuric acid were added and the funnel was shaken vigorously for 3 min. After the phases had been allowed to separate, the aqueous layer was run off and the benzene layer extracted with 25 mℓ of one percent w/w sodium carbonate solution, the mixture being shaken for 3 min and then allowed to settle. The aqueous layer was rejected and the benzene layer washed with 20-mℓ portions of distilled water until neutral to pH indicator paper. The benzene solution was dried over anhydrous magnesium sulphate and then decanted off. This solution is stable for several weeks if stored in the dark. (PEG 4000 did not form an ester under these reaction conditions).

	R_F	
Polyglycol 3,5-Dinitrobenzoate	**S1**	**S2**
PEG 200	0—0.8	0.5—0.8
PEG 300	0—0.8	0.3—0.8
PEG 600	0—0.7	0.2—0.8
PEG 800	0—0.2	0.1—0.3
PEG 1500	0—0.1	0.0—0.1
PPG 150	0.7—0.9	
PPG 425	0.5—0.8	
PPG 750	0.5—0.8	
PPG 1025	0.4—0.7	
PPG 1500	0.35—0.65	
PPG 2000	0.3—0.6	

Note: S1 = Ethyl acetate/acetic acid (97.5:2.5 v/v).
S2 = Water saturated methyl ethyl ketone.

Salvage, T., *Analyst*, 95, 363, 1970.

Table TLC-16
POLY(METHYLENE-2-HYDROXYBENZOIC ACID) FRACTIONATION

Layer	Silica gel H (E. Merck®) 0.75 mm
Amount	Polymer (0.3 g) in 2 mℓ acetone-water (10:1)
Detection	Ultraviolet
Solvent	Benzene/acetic acid/water (2:2 1)

Fraction	$R_F \times 100$	Equiv. wt.	Mol. wt.
81	0—4	158.8	1100
2	5—18	160.9	800
3	19—27	165.3	550
4	28—38	173.8	450
5	39—46	143.8	228[a]
PMSA	—	163.9	650
Salicylic acid	78	138.2	138

[a] This is the molecular weight of 3,3′-dicarboxy-4,4′-dioxydiphenylmethane.

Cimpa, C., Grieco, C., and Silipo, C., *J. Chromatogr.*, 46, 132, 1970.

Table TLC-17
POLY(OXYPROPYLENE DIOL) PRESENT IN POLY(OXYPROPYLENE TRIOL)

Layer	Silica 7G (Mallinckrodt®)
Detection	Sprayed with conc. H_2SO_4 — oven 110°C for 30 min quantitated by densitometry
Amount	1—2 μℓ (100—200 μg)
Solvent	Water saturated ethyl acetate

Standards of diol and triols totaling 1.0 ± 0.05g were weighed into 10-mℓ volumetric flasks and dissolved in chloroform. Samples were prepared in the same concentration range. A 1—2 μℓ aliquot of standards and samples were alternately spotted on the TLC plate and separated as below:

Avg. Mol. wt.	R_F diol	R_F triol
400	0.48	0.25
1000	0.44	0.24
2000	0.40	0.16
3000	0.37	0.12

Mameniskis, W. R., *J. Appl. Polym. Sci.*, 14, 1189, 1970.

Table TLC-18
POLYSTYRENE FRACTIONATION

Layer		L1	L2
Amount	10—20 μg polymer		
Detection	Fluorescence		
Solvent		S1	S2

Polystyrene MW	$\overline{M}w/\overline{M}w$	R_F	
10,300	1.06	1.0	0.81
19,800	1.06	0.78	0.69
51,000	1.06	0.48	0.46
160,000	1.06	0.12	0.39
411,000	1.06	0.03	0.33
360,000		0.00	0.29
1,800,000	1.20	0.00	0.19

Note: Layer — L1 = Silica gel F_{254} (Brinkmann Instruments®).
L2 = Alumina F_{254} (Brinkmann Instruments®).
S1 = Acetone-chloroform (4:1).
S2 = Gradient acetone to acetone-tetrahydrofuran (7:3).

Otocka, E. P. and Hellman, M. Y., *Macromolecules,* 3(3), 362, 1970.

Table TLC-19
MOLECULAR WEIGHT DISTRIBUTION OF POLYSTYRENE

Layer Silica gel, polyester backed and fluorescent indicator (Brinkmann Instruments®)
Detection Quenched fluorescence
Amount 10 g standards and 30 g samples applied in 0.4 × 1.0 cm streaks

Solvent	S1	S2	S3	
Fraction	$\overline{M}w/\overline{M}n$			
PS-2	19,800	1.06	0.44	0.90—0.95
PS-3	51,000	1.06	0.40	0.90
PS-4	160,000	1.06	0.33	0.85
PS-5	411,000	1.06	0.30	0.85
PS-6	860,000	1.10	0.27	0.80
PS-7	1,800,000	1.20	0.23	0.78

Note: Solvent — S1 = benzene/hexane (34:7).
S2 = dioxane/methanol (71.4:28.6).
S3 = benzene/isopropanol (64:2:35.8).

Otocka, E. P., *Macromolecules,* 3, 691, 1970.

Table TLC-20
POLYSTYRENE AND POLY(α-METHYLSTYRENE) OLIGOMERS

Layer	Silica gel KSK 5—10 μm	
Detection	Fluorescence	
Solvent	S1	S2

Polystyrene		Poly(α-methylstyrene)	
Mn	R_F	Fraction	R_F
314	0.90	tetramer	0.39—0.62
418	0.79	hexamer	0.17—0.46
600	0.05—0.90	octamer	0.04—0.32
900	0.00—0.79	decamer	0.00—0.18
2000	0.00—0.31		

Note: Solvent — S1 = Cyclohexane-benzene (14:3).
S2 = Carbon tetrachloride-heptane (2:1).

Belenky, B. G., Valchikhina, M. D., Vakhtina, I. A., Gankina, E. S., and Tarakanov, O. G., *J. Chromatogr.*, 129, 115, 1976.

Table TLC-21
STYRENE-BUTADIENE COPOLYMERS

Layer	L1	L2
Detection	Thymol blue method	
Amount	1—2 μℓ 0.5g polymer 100 mℓ carbon tetrachloride	
Solvent	As listed in table	

			R_F for L1					R_F for L2				
Solvent	Dielec. Const.	ε°	PS	PB	Tap-SB	Di-	Tri-	PS	PB	Tap-SB	Di-	Tri-SB
Cyclohexane	2.02	0.04	0	0	0	0	0	0	0	0	0	0
CCl_4	2.24	0.18	0	1	0	0	0	0	1	1	1	0
Amyl chloride	6.6	0.26	0	1	1	1	1	1	1	1	1	1
p-Xylene	2.27	0.26	1	1	1	1	1	1	1	1	1	1
Benzene	2.28	0.32	1	1	1	1	1	1	1	1	1	1
$CHCl_3$	4.31	0.40	1	1	1	1	1	1	1	1	1	1
THF	7.42	0.45	1	1	1	1	1	1	1	1	1	1

Note: ε° = Solvent strength parameter defined by Snyder.
R_F changes from zero to unity according to the microstructure.

PS = polystyrene.
PB = polybutadiene.
Tap-SP = tapered styrene-butadiene copolymer.
Di- = styrene-butadiene-block.
Tri- = styrene-butadiene-styrene block.
Layer — L1 = Silica gel B-O; 0.5 mm (Wako Chemical Co., Tokyo)®.
L2 = Aluminum oxide G; 0.5 mm (E. Merck)®.

DonKai, N., Miyamoto, T., and Inagaki, H., *Polym. J.*, 7, 577, 1975.

Table TLC-22
UREA-FORMALDEHYDE-METHANOL REACTION PRODUCTS

Layer — Kieselgel F_{254} (Merck)® (exposed to ammonium vapor)
Detection — After air drying the plate for 15 min, expose to acetic acid solution
Amount — 1 μℓ of 5% in methanol
Solvent

		S1	S2
		R_F	
Abbreviation	Compound		
DMU.DME	dimethylolurea, DME	.88	.78
MMU.MME	monomethylolurea, MME	.84	.61
DM.MDU.DME	dimethylolmethylenediurea DME	.84	.55
DMU.MME	dimethylolurea MME	.67	.48
MM.MDU.MME	monoethylolmethylene diurea MME	.66	.38
Urea	Urea	.65	.38
Formaldehyde	Formaldehyde	.57	.35
DM.MDU.MME	dimethylolmethylenediurea MME	.55	.26
MMU	monomethylolurea	.50	.24
MDU	methylenediurea	.50	.24
DMU	dimethylolurea	.32	.15
MM.MDU	monomethylolmethylenediurea	.32	.13
DM.MDU	dimethylolmethylenediurea	.18	.08

Note: MME = monomethyl ether.
DME = dimethyl ether.

Solvent — S1 = Ethylacetate-methanol-ammonia solution (80:15:5).
S2 = Ethyl acetate-methanol-ammonia solution (88:6:6).

Ludlam, P. R., *Analyst,* 98, 107, 1973.

Section V

Compound Index

INDEX

A

B

C

D

E